I0502974

About the National Science and Technology Council

The National Science and Technology Council (NSTC) was established by Executive Order on November 23, 1993. This cabinet-level council is the principal means by which the President coordinates science, space, and technology policies across the Federal Government. NSTC acts as a virtual agency for science and technology to coordinate diverse paths of the Federal research and development enterprise. An important objective of the NSTC is the establishment of clear national goals for Federal science and technology investments in areas ranging from information technologies and health research to improving transportation systems and strengthening fundamental research. The Council prepares research and development strategies that are coordinated across the Federal agencies to form a comprehensive investment package aimed at accomplishing multiple national goals. For additional information about the NSTC, please visit our website at http://www.ostp.gov/nstc/html/NSTC_Home.html.

About the Office of Science and Technology Policy

The Office of Science and Technology Policy (OSTP) was established by the National Science and Technology Policy, Organization and Priorities Act of 1976. OSTP's responsibilities include advising the President in policy formulation and budget development on all questions in which S&T are important elements; articulating the President's S&T policies and programs; and fostering strong partnerships among Federal, state and local governments, and the scientific communities in industry and academe. For additional information about OSTP, please visit our website at http://www.ostp.gov.

About this Report

This report, prepared by the NSTC Subcommittee on Biometrics and Identity Management, highlights key US Government initiatives in advancing the science of biometrics and its utilization in meeting pressing operational needs. While federal efforts in biometric technologies predate 9/11/2001 by several decades, this report primarily focuses on the breath and impact of the significantly enhanced attention and progress made since that date.

Acknowledgements

The NSTC Subcommittee on Biometrics and Identity Management would like to thank our partners from the academic, industrial, and privacy communities for their guidance and assistance over the past seven years. Our scientific and operational successes would not have been possible without this partnership.

The Subcommittee offers special thanks to the following Subcommittee members for authoring portions of this paper:

Duane Blackburn, EOP/OSTP
Tom Coty, DHS/S&T
John Cook, NCC
Tom Dee, DoD/ATL
Jeff Dunn, NSA
Patrick Grother, DOC/NIST
Mike Hogan, DOC/NIST

Mike Garris, DOC/NIST
Michael King, IC
Michael Lilienthal, DoD/BTF
Ross Michaels, DOC/NIST
Chris Miles, DHS/S&T
Elaine Newton, DOC/NIST
Fernando Podio, DOC/NIST

Jonathon Phillips, DOC/NIST
Scott Swann, FBI/CJIS
Elham Tabassi, DOC/NIST
Mary Theofanos, DOC/NIST
Kimberly Weissman, DHS/US VISIT

Finally, the Subcommittee would like to thank Heather Rosenker and Megan Hirshey, US-VISIT contractors, for their editorial assistance, as well as the FBI CJIS Division for providing graphics, editing, publishing and printing services in support of this paper.

September 11, 2008

Dear Colleagues:

This report, prepared by the National Science and Technology Council (NSTC) Subcommittee on Biometrics and Identity Management, documents key US Government initiatives to advance the science of biometrics and assesses their value in meeting critical operational needs. While federal efforts in biometric technologies predate the terrorist attacks of September 2001 by several decades, this report focuses on progress made since then.

Working through the NSTC, and in cooperation with the academic and industrial research communities, agencies embarked on a multi-year initiative to advance the capabilities of biometric technologies. As capabilities advanced, agencies quickly incorporated them into their operational systems and then worked to develop government-wide policies on how to use biometrics to support missions against known and suspected terrorists, while simultaneously enhancing privacy protection for US Citizens and foreign visitors.

By developing a common planning focus for departments and agencies we have advanced the technology and its operational implementation at a far greater pace than would have been possible otherwise. Today, federal agencies are using biometrics to enhance security and operational efficiency throughout the nation, at the borders and in the battlefields of Afghanistan and Iraq. Their continued efforts to meet the ongoing needs outlined in *The National Biometrics Challenge* will ensure even greater successes in the future.

Sincerely,

John H. Marburger, III
Director

TABLE OF CONTENTS

Introduction

On September 11, 2001, 19 terrorists boarded aircraft in Boston, Mass., and Dulles, Va., and changed our world. All had successfully passed through security screening prior to boarding the aircraft and, previously, had also successfully passed through immigration screening while entering the country. A suspected 20th terrorist had been refused entry by a suspicious immigration inspector at Florida's Orlando International Airport the previous month. Of the remaining 19 terrorists, 18 had been issued U.S. identification documents. The global war on terror had reached American soil, and the terrorists had already realized how important identity was to be in this fight.

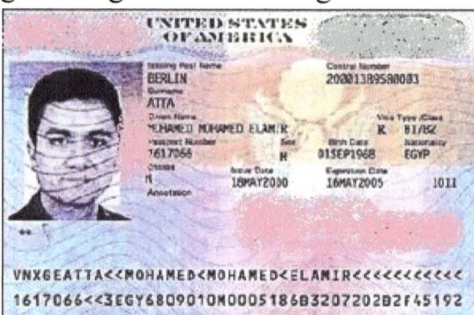

> **"Sources of identification are the last opportunity to ensure that people are who they say they are and to check whether they are terrorists."**
>
> **"For terrorists, travel documents are as important as weapons."**
>
> *--The 9/11 Commission Report*

While these terrible events were unfolding, a group of approximately 30 individuals from government, industry, and academia were in a hotel conference room in Orlando, Fla., at a Biometric Interoperability, Performance and Assurance Working Group[1] meeting. Cell phones and beepers started going off, alerting the workshop attendees of the ongoing terrorist attacks. The thoughts of those present were initially the same as those of any other citizen: "What is going on? Do I know anyone that may be hurt?"

These individuals were participating in a pre-conference workshop of the Biometric Consortium, an interagency body to discuss and coordinate biometric activities within the federal government. The conference that was to have started the next day would have been a small affair consisting of approximately three hundred technologists working on the bleeding edge of a nascent technology. The agenda for that conference did not include any presentations on active or planned government biometric systems. This small group knew that the world had instantly changed and that their tiny world of biometrics was about to experience significant change, as it would soon be called upon to help enhance security in many facets of government business. The group did not yet grasp the extent of this calling, nor how unprepared the community was to meet it.

---

[1] A working group sponsored by the National Institute of Standards and Technology (NIST) and the Biometric Consortium.

Initial Reactions

Biometrics
A general term used alternatively to describe a characteristic or a process.

As a characteristic:
A measurable biological (anatomical and physiological) and behavioral characteristic that can be used for automated recognition.

As a process:
Automated methods of recognizing an individual based on measurable biological (anatomical and physiological) and behavioral characteristics.

Biometrics was still very much a nascent technology and had not had a chance to properly mature before being thrust into the national spotlight. Cross-mission biometric standards were practically non-existent. The science of biometric testing was very much in its infancy with only one open, statistically relevant, evaluation of commercial biometrics having been performed. Government, industry, and academia had little experience working collaboratively within each sector, much less across sectors. The media had virtually no knowledge of biometric technologies and issues, and those with this knowledge had little experience working with the media. The hole was filled by instant experts, which led to numerous inaccurate press articles. Numerous entities with little to no understanding of biometrics technologies, of how to introduce users to the technology, or of how to ensure privacy, nonetheless rushed to be the first to install them. The results were predictable and established a feeling of distrust of the technology throughout the country that still exists to some degree today.

The Federal Aviation Administration (FAA) quickly established, with support and guidance from the Department of Defense, the Aviation Security Biometrics Working Group (ASBWG) to perform an initial analysis of the efficacy of integrating biometrics into airport security systems. The rapid (less than two months) work of this *ad hoc* group not only provided a reality check for the FAA and a foundation for biometric decisions in the soon-to-be-created Transportation Security Administration (TSA), but also identified several difficult issues that would need dedicated attention by government agencies for some time to come. Indeed, although aviation security was the primary focus for biometric applications in the immediate aftermath of 9/11, the possibilities for applying biometrics to homeland security and counterterrorism were quite broad. Federal agencies faced an unenviable task of having limited in-house biometrics expertise while simultaneously managing five critical activities:

- Rapidly integrating existing biometric capabilities into operational systems to meet critical short-term needs;
- Advancing technology so that future systems better met long-term operational needs;
- Advancing privacy theory and applying it to biometric activities;
- Overcoming technical, policy, and interagency trust issues to transition from traditional, stove-piped operational systems to government-wide interoperability of systems to meet counter-terrorism needs;

- Educating government officials and the citizenry about biometric technologies, their capabilities and limitations, and how they should be used.

This report examines how these five key activities have been addressed since 2001 across four operational areas: immigration and border management, law enforcement, intelligence and counterterrorism, and access control and credentialing. Although this report breaks out biometric activities by agency, this does not mean that each agency's biometric initiatives have been developed in a vacuum. On the contrary, agencies have laudably avoided stovepipes in order both to drive innovation and to achieve interoperability. A close reading of this report will reveal that cross-agency collaboration on biometric initiatives has been significant and directly contributed both to advancing biometric science and to enhancing federal operations.

General Timeline of Federal Government Biometric Activities

> **Key**
> - **Policy, Legislation, and General Events**
> - **Research, Development, Testing and Evaluation; Standards**
> - **Operations**

1967 — The Federal Bureau of Investigation (FBI) and NIST begin research on technologies for the automated matching of fingerprints.

1986 — ANSI/NBS-ICST 1-1986 "American National Standard – for Information Systems – Fingerprint Identification – Data Format for Information Interchange" is adopted.

1992 — The Biometric Consortium is established within the U.S. government.

1993 — The Department of Defense (DoD) initiates the FacE REcognition Technology (FERET) program.

— Immigration and Naturalization Service (INS) initiates INSPASS using hand geometry at ports of entry to facilitate the inspection of business travelers to the US (using special kiosks and lanes that bypass the normal inspection lanes)

1994 — FBI plans development of the Integrated Automated Fingerprint Identification System (IAFIS).

— INS' IDENT system becomes operational.

1995 — Iris prototype becomes available as a commercial product.

— INS uses facial recognition and voice recognition to verify the identity of pre-enrolled persons in vehicles crossing the border at Otay Mesa, California using a special, dedicated lane.

1996 — The Illegal Immigration Reform and Immigrant Responsibility Act of 1996 becomes law.

— NIST begins hosting annual speaker recognition evaluations.

— INS opens the first fully automated Port of Entry at Scobey, Montana relying upon voice verification technology

1997 — The Human Authentication Application Program Interface (API), the first commercial, generic, biometric interoperability standard is published.

1998 — The Department of State (DOS) begins collecting biometrics from Mexican nationals applying to enter the United States with a Border Crossing Card (BCC).

1999 — FBI's IAFIS major components become operational.

2000 — The first Face Recognition Vendor Test (FRVT 2000) is held.

— DoD establishes its Biometrics Management Office (BMO) and Biometrics Fusion Center (BFC).

— The Visa Waiver Program Act of 2000 becomes law.

— The Defense Advance Research Projects Agency (DARPA) begins the Human Identification at a Distance (HumanID) Program.

— The Biometric Application Programming Interface (BioAPI) specification is released.

2001 — The terrorist attacks of September 11, 2001 occur.

— FAA establishes Aviation Security Biometrics Working Group.

— INCITS establishes M1 Technical Committee on Biometrics.

— The USA PATRIOT Act becomes law.

— The Center for Identification Technology Research (CITeR) begins operation as a National Science Foundation Industry/University Cooperative Research Center.

2002 — The Enhanced Border Security and Visa Entry Reform Act becomes law.

— The ISO/IEC SC 37 standards subcommittee on biometrics is established.

— The Maritime Transportation Security Act becomes law, establishing the Transportation Worker Identification Credential (TWIC).

— FRVT 2002 is held.

— The E-Government Act of 2002 becomes law.

2003 — DOS begins collecting biometrics from visa applicants through the BioVisa program.

— The National Science and Technology Council charters a Subcommittee on Biometrics to coordinate biometrics research and development (R&D), policy, outreach, and international collaboration across the federal government.

— ICAO adopts blueprint to integrate biometrics into machine-readable travel documents.

— The Department of Justice (DOJ), DOS and NIST submit joint Patriot Act report to Congress on "Use of Technology Standards and Interoperable Databases with Machine-Readable, Tamper-Resistant Travel Documents"

— Testing for the Fingerprint Vendor Technology Evaluation (FpVTE 2003) begins.

— NIST begins Proprietary Fingerprint Template (PFT) testing.

— Homeland Security Presidential Directive (HSPD) 6 establishes the Terrorist Screening Center.

2004 — HSPD-11 establishes a coordinated and comprehensive approach to terrorist-related screening.

— HSPD-12 calls for standard, government-wide personal identification verification (PIV) credentials for all federal employees and contractors.

— Face Recognition Grand Challenge begins.

— International Meeting of Biometrics Experts held.

— DHS' US-VISIT program begins collecting biometrics from international visitors at all international air, sea, and land border ports of entry.

— AirNexus kickoff – facilitated travel program operated jointly with the Government of Canada using iris verification at kiosks for pre-enrolled travelers

- Slap Fingerprint Segmentation Evaluation 2004 (SlapSeg04).
- DoD's IAFIS-compatible database, Automated Biometric Identification System (ABIS), becomes operational.
- The fingerprint Minutiae Interoperability Exchange 2004 (MINEX 04) tests begin.
- The Intelligence Reform and Terrorism Prevention Act becomes law.
- DHS and NIST initiate a program to develop human computer interaction (HCI) guidelines and standards for biometric systems.
- The first statewide automated palm print databases in the United States are deployed in California, Connecticut, and Rhode Island.
- The first Common Biometric Exchange Formats Framework (CBEFF) ANSI standard is published.
- NIST Fingerprint Image Quality assessment tool is released.

2005
- NIST issues standards for federally mandated, government-wide PIV cards.
- NIST hosts the 10-Print Capture Scanner & Software Requirements Workshop.
- The Iris Challenge Evaluation 2005 (ICE 2005) Program is held.
- The European Commission hosts a "Workshop on Ethical and Social Implications of Biometric Identification Technology: Toward an International Approach."

2006
- DoD reorganizes the BMO and the BFC into the Biometrics Task Force (BTF).
- FRVT 2006 is held.
- First international *NIST Biometric Quality Workshop* is held.
- Defense Science Board launches Task Force to study biometrics in the DoD.
- DHS hosts the 10-Print Capture User Group Industry Day.
- NIST holds the Latent Fingerprint Testing Workshop.
- Agencies, working through the NSTC, begin the process of designing government-wide biometric system interoperability.
- The President approves the *National Implementation Plan for the War on Terror.*
- www.biometrics.gov is launched.
- *The National Biometrics Challenge* is issued.
- ICE 2006 is held.
- The United States hosts the "International Conference on Biometrics and Ethics"

2007
- Agencies, working through the NSTC and National Counterterrorism Center, begin collaboration to improve the coordination of biometric activities to support efforts against known and suspected terrorists.
- TWIC Enrollment and Issuance begins.
- Technology demonstrations for the Fast Capture Rolled-Equivalent Finger/Palm Print Initiative begin.
- NSTC expands the focus of its existing biometrics subcommittee, creating the NSTC Subcommittee on Biometrics and Identity Management.
- NIST conducts Phase I of the Evaluation of Latent Fingerprint Technologies (ELFT).
- ANSI/NIST-ITL 1-2007 – Data Format for the Interchange of Fingerprint, Facial, & Other Biometric Information – Part 1 is adopted.
- NIST conducts MINEX II (Fingerprint Match on Card).
- DHS *Privacy Technology Implementation Guide* is issued.
- *NSTC Policy for Enabling the Development, Adoption and Use of Biometric Standards* is issued.

2008

— NIST releases public domain build instructions for the Multimodal Biometric Application Resource Kit (MBARK).

— DOS begins deploying 10-fingerprint collection at all visa-issuing posts.

— US-VISIT begins deploying 10-fingerprint collection at all U.S. airports.

— The NSTC Task Force on Identity Management is chartered.

— The Multiple Biometric Grand Challenge (MBGC) begins.

— DHS begins accepting applications for the Global Entry expedited trusted traveler program.

— FBI plans development of the Next Generation Identification System to incorporate multimodal biometrics.

— NSTC *Registry of USG Recommended Biometric Standards* is issued.

— The President issues NSPD-59/HSPD-24: *Biometrics for Identification and Screening to Enhance National Security*.

— The International Workshop on Usability and Biometrics is held.

— ANSI/NIST-ITL 2-2008 – XML Data Format for the Interchange of Fingerprint, Facial & Other Biometric Information – Part 2 is adopted.

Advancing the Science of Biometrics

To ensure that U.S. biometrics-based systems would meet the government's long-term operational needs, the U.S. government recognized the need to improve its understanding of biometrics and lead the effort to advance the technology's capabilities. After 9/11, the federal government initiated a series of activities focused on research, development, testing, and evaluation (RDT&E), as well as standards development. These activities were collaboratively planned, funded, and managed by multiple federal agencies, as coordinated through the NSTC Subcommittee on Biometrics and Identity Management. Scientists at the National Institute of Standards and Technology (NIST) often provided technical leadership and performed day to day management of the individual projects.

These activities have enabled the U.S. government to establish new, and enhance existing, biometrics-based systems that are both improving the security of the United States and maintaining personal privacy and civil liberties. The following section of this report addresses the U.S. government's progress to date on RDT&E and development of biometric standards.

RDT&E and Standards Timeline

1993 — DoD initiates the FacE REcognition Technology (FERET) program.

1996 — NIST begins hosting annual speaker recognition evaluations.

1997 — The Human Authentication API, the first commercial, generic biometric interoperability standard, is published.

2000 — The first Face Recognition Vendor Test (FRVT 2000) is held.

— The Defense Advance Research Projects Agency (DARPA) begins the Human Identification at a Distance (HumanID) Program.

— The Biometric Application Programming Interface (BioAPI) specification is released.

2001 — The Center for Identification Technology Research (CITeR) begins operation as a National Science Foundation Industry/University Cooperative Research Center.

— INCITS establishes M1 Technical Committee on Biometrics.

2002 — The ISO/IEC SC 37 standards subcommittee on biometrics is established.

— FRVT 2002 is held.

2003 — The NSTC charters a Subcommittee on Biometrics to coordinate biometrics R&D, policy, outreach, and international collaboration across the federal government.

— Testing for the Fingerprint Vendor Technology Evaluation (FpVTE 2003) begins.

— NIST begins Proprietary Fingerprint Template (PFT) testing.

2004 — Face Recognition Grand Challenge begins.

— Slap Fingerprint Segmentation Evaluation 2004 (SlapSeg04).

— The fingerprint Minutiae Interoperability Exchange 2004 (MINEX 04) tests begin.

— DHS and NIST initiate a program to develop human computer interaction (HCI) guidelines and standards for biometric systems.

— The first Common Biometric Exchange Formats Framework (CBEFF) ANSI standard

2005
is published.
— NIST Fingerprint Image Quality assessment tool is released.
— NIST issues standards for federally mandated, government-wide PIV cards.
— NIST hosts the 10-Print Capture Scanner & Software Requirements Workshop.
— The Iris Challenge Evaluation 2005 (ICE 2005) Program is held.

2006
— FRVT 2006 is held.
— The first international *NIST Biometric Quality Workshop* is held.
— DHS hosts the 10-Print Capture User Group Industry Day.
— NIST holds the Latent Fingerprint Testing Workshop.
— *The National Biometrics Challenge* is issued.
— ICE 2006 is held.

2007
— Technology demonstrations for the Fast Capture Rolled-Equivalent Finger/Palm Print Initiative begin.
— NIST conducts Phase I of the Evaluation of Latent Fingerprint Technologies (ELFT).
— ANSI/NIST-ITL 1-2007 – Data Format for the Interchange of Fingerprint, Facial, & Other Biometric Information – Part 1 is adopted.
— NIST conducts MINEX II (Fingerprint Match on Card).
— *NSTC Policy for Enabling the Development, Adoption and Use of Biometric Standards* is issued.
— NIST releases public domain build instructions for the Multimodal Biometric Application Resource Kit (MBARK).

2008
— The Multiple Biometric Grand Challenge (MBGC) begins.
— NSTC Registry of USG Recommended Biometric Standards is issued.
— The International Workshop on Usability and Biometrics is held.
— ANSI/NIST-ITL 2-2008 – XML Data Format for the Interchange of Fingerprint, Facial & Other Biometric Information – Part 2 is adopted.

1. Research, Development, Test & Evaluation (RDT&E)

After 9/11, the U.S. government has primarily focused its RDT&E efforts on four biometrics—face, finger, iris, and multimodal. Agencies throughout the U.S. government, working with partners from industry and academia, have contributed to these efforts. All federal biometrics RDT&E is closely prioritized and coordinated through the NSTC Subcommittee on Biometrics and Identity Management and often involves joint sponsorship and project management from multiple agencies. Prioritized RDT&E needs are described in *The National Biometrics Challenge*. A brief summary of RDT&E efforts is provided in this section. Additional detail about each effort is provided in Appendix A.

a. Face Recognition

As face recognition technology became commercialized in the late 1990s, the government needed a way to measure the performance of these systems in a statistically relevant manner. In 2000, the U.S. government began a series of evaluations for face recognition systems, known as the Face Recognition Vendor Tests (FRVT), which were continued after the attacks of 9/11. There have been three FRVT evaluations since 2000. Each successive evaluation increased in size, difficulty, and complexity. These evaluations not only provided snapshots in time of face recognition capabilities, but also drove continuing advancement of the technology world-wide. The most recent FRVT results also showed that several automatic face recognition algorithms were comparable to or better than humans at recognizing faces taken under different lighting conditions.

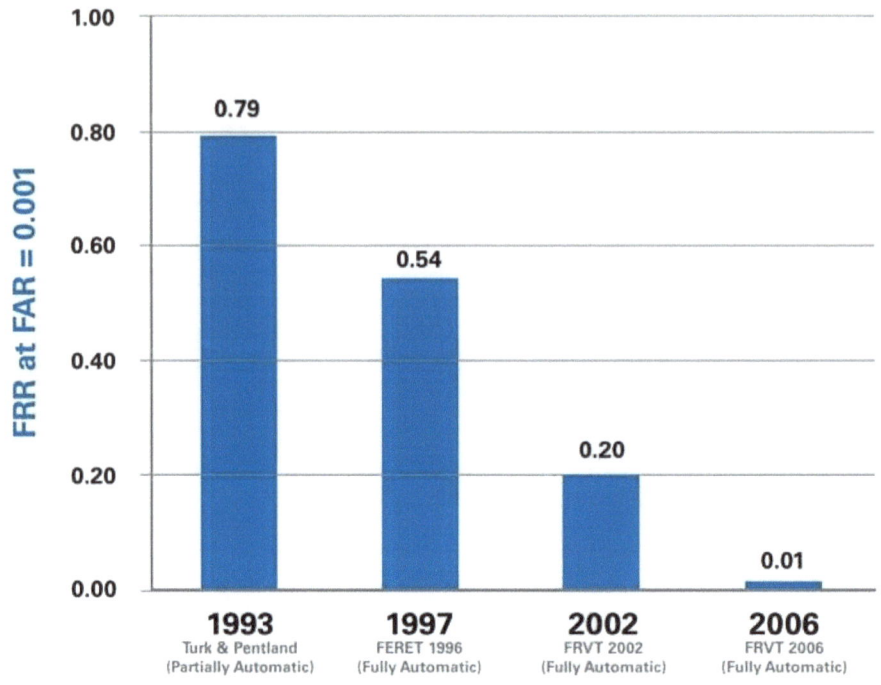

The reduction in error rate for state-of-the-art face recognition algorithms as documented through the FERET, the FRVT 2002, and the FRVT 2006 evaluations.

The Face Recognition Grand Challenge (FRGC) was initiated in 2004 to improve face recognition verification performance by an order of magnitude over the FRVT 2002 results. Challenge problems were developed with a set of experiments designed to guide technology development to meet U.S. government operational requirements.

The government also sponsored the Face Recognition Advanced Study Workshop in 2005. The purpose of the workshop was twofold: to discuss advancement hurdles, recent seminal works related to those hurdles, and ideas for future research topics; and to stimulate research and cross-institution collaboration among the most promising young scientists in the maturing field. A total of 55 individuals participated in this invitation-only workshop, where participants were generally sequestered for two and a half days of intensive technical deliberations.

The workshop's format was similar to defending a PhD dissertation before the nation's recognized experts and produced highly interactive discussions.

b. Fingerprint Identification and Verification

Since 2001, the U.S. government has increased its efforts to advance and evaluate fingerprint recognition technology. The government has conducted vendor technology evaluations, Proprietary Fingerprint Template (PFT) testing, slap fingerprint segmentation evaluations, fast fingerprint slap and rolled-equivalent capture, latent fingerprint testing, and fingerprint minutiae interoperability testing.

The U.S. government began the Fingerprint Technology Vendor Evaluation (FpTVE) in 2003. It was the most comprehensive independent evaluation of fingerprint matching systems ever executed, particularly in terms of the number and variety of systems and fingerprints. From this testing, the U.S. government learned that top-performing systems performed consistently well over a variety of image types and data sources, and they produced matching accuracy results that were substantially different from less robust systems. The testing also statistically confirmed the degree to which the use of additional fingers improve accuracy, as well as the degradation of accuracy caused by the collection of poor-quality fingerprints. Additionally, this test showed that fingerprint scanners alone do not determine fingerprint image quality.

Through Proprietary Fingerprint Template (PFT) testing, the U.S. government continues to test fingerprint-based biometric matching systems using vendor-supplied software development kits. The PFTs measure the state-of-the-art in one-to-one matching for verification over a wide range of fingerprint image qualities. This testing is important to ensure that the fingerprint matching algorithms being used in existing and planned government systems are state of the art.

To assess the accuracy of algorithms used to segment slap fingerprint images into individual fingerprints, the U.S. government has conducted Slap Fingerprint Segmentation Evaluations since 2004. In the first Slap Fingerprint Segmentation test, SlapSeg 04, the U.S. government evaluated segmentation algorithms on a variety of operational-quality slap fingerprints based on their abilities to produce highly matchable images, identify finger positions, and detect segmentation failures.

Slap Fingerprint

Fingerprints taken by simultaneously pressing the four fingers of one hand onto a scanner or a fingerprint card, as illustrated below. Slaps are known as four finger simultaneous plain impressions.

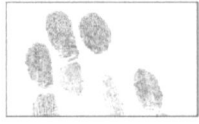

Segmentation

The process of parsing the biometric signal of interest from the entire acquired data system. For example, finding individual finger images from a slap impression, as illustrated below.

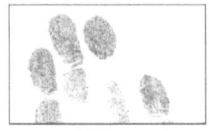 

In particular, information obtained through SlapSeg 04 has played a critical role in the work of the Department of Justice (DOJ) and the Department of Homeland Security (DHS) to make their fingerprint databases, IAFIS and IDENT respectively, interoperable. The National Institute of Standards and Technology (NIST) plans a new series of segmentation evaluations called SlapSeg II that will provide the U.S. government with information about how the current state of the art in slap segmentation has advanced.

The U.S. government has also worked to advance both rolled-equivalent and slap capture devices. In 2004, the U.S. government began the Fast Capture Rolled-Equivalent Finger/Palm Print Initiative, a multi-year, applied research program. The goal of this effort is to enable the ability to capture rolled-equivalent fingerprints in 15 seconds or less and both palms in 1 minute or less. In 2005, agencies across the U.S. government identified joint needs for faster, smaller, slap capture 10-fingerprint scanners. The agencies issued a Request for Information and held two industry days at which they set their operational requirements and then refined their operational requirements after industry's initial response with fingerprint scanners. Today, the slap capture 10-fingerprint scanners procured through this process are being used at U.S. visa-issuing posts, ports of entry, and for civilian background checks.

A U.S. Customs and Border Protection officer demonstrates how an international visitor should place her fingers on the scanner during the entry process.

The federal government is also conducting a series of tests to evaluate the state of the art in automated latent fingerprint matching, called Evaluation of Latent Fingerprint Technologies (ELFT). ELFT is structured as a multi-year project. The first part of this project consists of two phases running in a "lights-out" environment. Phase I was completed in 2007 and represents a proof-of-concept test whose main purpose was to demonstrate integrity of the software, including the evaluation of the test-bed itself. Phase II is currently under way and employs a larger database to quantify the achievable performance ("hit rate") for automated searches.

Finally, the federal government established the Minutiae Interoperability Exchange (MINEX) program to determine the feasibility of using minutiae data (rather than image data) as the interchange medium for fingerprint information between different fingerprint matching systems. This program is made up of three tests with a fourth test planned, and the results from these tests have been influential on various biometric-based identity management programs. These tests have helped define the structure of an identity credential, established a test to which vendors can submit their products to ensure PIV compliance, and demonstrated the use of match-on-card verification algorithms as a means of privacy enhancement.

c. Iris Recognition

The U.S. government funded iris recognition research for several years prior to 9/11. While iris recognition technology garnered acceptance as a highly accurate biometric, it required a high degree of cooperation from the user. To improve the utility, performance, and ease-of-use of this technology, the U.S. government substantially increased its investment after 9/11. Notable advancements that can be attributed to this investment and foresight include but are not limited to: increased standoff distances, increased system performance (while reducing size and cost), and the demonstration of prototypes capable of acquiring and matching the iris of subjects while moving through a portal. In addition, the U.S. government has sponsored the development of multiple matching algorithms, including government-owned algorithms.

Other areas of influence include the sponsorship of academic programs to create U.S. experts and spawn new technologies that encourage commercial competition and foster the rapid introduction of technological advancements. In addition, the U.S. government has sponsored the development of multiple analysts' tools that augment automated iris match algorithms to address the needs of a broad array of government, industry, and academic partners. These activities have significantly advanced the state of the art and enabled interoperable iris biometric technology.

d. Multimodal Biometric Identification

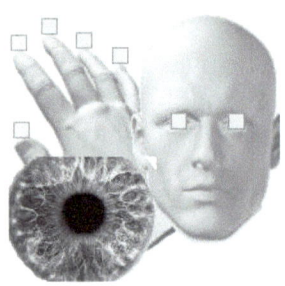

Prior to 9/11, the U.S. government had already begun efforts to develop automated, multimodal systems for identifying people at a distance for protection and early warnings against asymmetric threats. One such effort was the Defense Advance Research Projects Agency's (DARPA) Human Identification at a Distance (HumanID) Program, which began in September 2000. The state-of-the-art capability at that time on cooperative subjects, indoors, with controlled illumination was less than 10 feet. Various types of biometric technology were explored, to include face recognition, iris recognition, Doppler radar, infrared imagery, pulse and heartbeat, and gait (recognizing someone by their walk). By the end of the program in 2003, some technology had improved from being able to recognize people at less than 10 feet to being capable of recognizing people at up to 150 feet. Overall, the HumanID program made significant gains in understanding the difficulties associated with biometric technology and provided the ground work for numerous future biometric research programs.

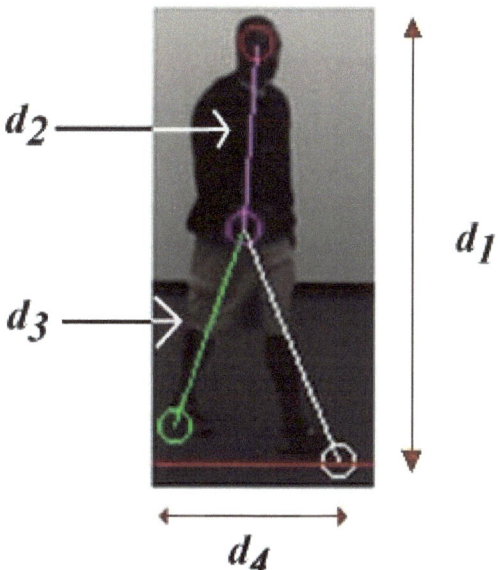

Today, the U.S. government is managing the Multiple Biometric Grand Challenge (MBGC) to foster theory and systems that can smartly use multiple biometric technologies. The MBGC's first set of results are planned for early 2009. Additionally, the U.S. government developed the Multimodal Biometric Application Resource Kit (MBARK). MBARK is public domain source code that provides a consistent and usability-tested user interface, which means that operators can more quickly recover from both minor mistakes and major hardware failures.

e. **Biometric Quality**

Results from as early as FpVTE 2003 clearly demonstrate that one of the most significant factors affecting biometric accuracy is that of quality. Test and evaluations demonstrate time and again that many algorithms perform well on high-quality biometric samples, but what separates the crowd is how algorithms perform on poor-quality samples. An effective quality measure can have many uses, but circa 2003, no publicly open or standard quality metrics existed. To address this gap, the federal government created a Biometrics Quality Program.

If quality can be improved, either by sensor design, user interface design, or standards compliance, better performance can be realized. For those aspects of quality that cannot be designed-in, an ability to analyze the quality of a live sample is needed. This is useful primarily in initiating the reacquisition from a user, but also for the real-time selection of the best sample and the selective invocation of different processing methods. That is why quality measurement algorithms are increasingly deployed in operational biometric systems. With the increase in deployment of quality algorithms, the need to standardize an interoperable way to store and exchange biometric quality scores increases.

Since 2004, the federal government has focused on standards, tools, guidance, and workshops.

- Standards. The federal government actively participates in SC 37 and M1 quality standardization activities, including making significant contributions to the draft ISO/IEC 2974 standard.
- Tools. The federal government released the NIST Fingerprint Image Quality (NFIQ) algorithm in August of 2004. NFIQ is a fingerprint quality measurement tool. It is implemented as open-source software and is used today in U.S. government and commercial deployments. Its key innovation is to produce a quality value from a fingerprint image that is directly predictive of expected matching performance, and it has been designed to be matcher independent. There is now international consensus in industry, academia, and government that a statement of a biometric sample's quality should be related to its recognition performance. Since its release, NFIQ has won national and international acceptance and has become a *de facto* standard.
- Guidance. NIST has published a technical contribution and guidance toward quality summarization, examined methods of assessing how effective a quality algorithm is in predicting performance, and conducted studies on incorporating quality in multimodal biometric systems.
- Workshops. The federal government has hosted a series of international Biometric Quality Workshops in March 2006 and November 2007 to discuss capabilities vis-à-vis operational requirements and to identify research needs, testing requirements, and standardization gaps. The workshops provided a forum for experts to share their research and discuss problems and new developments.

f. Biometrics Usability

A more recent avenue of scientific research is the human computer interaction (HCI) of biometric systems. The federal government recognized this need and initiated a program in 2004 to develop HCI guidelines and standards for biometric systems.

The goal of the usability effort is the development and testing of a set of usability guidelines for biometric systems that enhance performance (throughput and quality), improve user satisfaction and acceptance, and provide consistency across biometric system user interfaces. Achieving these goals requires an understanding of the users, user behavior, and the biometric systems.

Fingerprint scanner height and angle have a direct influence on image capture, and thus, operational success. The federal government analyzed this issue, and developed recommendations and adjustable mounting brackets to maximize collection quality over a diverse population. US-VISIT quickly implemented the results of this research into their operational collection sites.

Six usability research studies have been conducted including the study of the impact of the following.

- user habituation or acclimatization
- counter height and anthropometrics
- instructional materials
- adaptable devices for accessibility
- international symbols
- relationship of counter height and angle of fingerprint scanners
- face overlays

These research studies have resulted in seven reports and two ISO standards submissions. These documents provide guidelines for implementation and deployment of biometric applications. The International Workshop on Usability and Biometrics was held in June of 2008 to further promote biometric usability studies. The test results have had a direct impact on existing and planned biometric deployments within biometrics programs, such as US-VISIT.

During the entry inspection process, international visitors have their fingers digitally scanned and a photograph taken. This process takes only seconds and is easy to do.

2. Development of Biometric Standards

In addition to studying various biometric technologies, the U.S. government has invested significant efforts into the development of biometric standards. While some very successful biometric standards activities had existed well before 2001, 9/11 provided an impetus to greatly expand and accelerate comprehensive standards development, as envisioned government and private sector systems required a solid standards base. The first step was to create formal working/technical groups in accredited standards development organizations to develop generic biometric standards that would support both identification and verification applications. The U.S. government spearheaded this effort by formally proposing these groups at the national (INCITS – InterNational Committee for Information Technology Standards) and international (Joint Technical Committee 1 of ISO/IEC) levels in October 2001 and January 2002, respectively. To further support these efforts, the U.S. government also assigned personnel and fiscal resources to lead these efforts. Since that time, 22 national standards and 25 international standards have been developed and approved. Several of these standards are now in their second versions.

By 2007, multiple competing versions of some standards existed[2]. To help ensure interoperability of government systems, the NSTC Subcommittee on Biometrics and Identity Management led an interagency effort to develop the *NSTC Policy for Enabling the Development, Adoption and Use of Biometric Standards* (September 2007). The goal of this policy is to establish a framework to reach interagency consensus on biometric standards adoption for the federal government and resulted in the release of the *Registry of US Government Recommended Biometric Standards* (June 2008). Federal agency adoption of these recommended standards and associated conformity assessment programs will enable necessary next generation federal biometric systems, facilitate biometric system interoperability, and enhance the effectiveness of biometric products and processes.

---

[2] This is not uncommon in the standards realm, as the national and international standards bodies work at different paces.

Operational Activities

The use of biometrics by government agencies to enhance operational capabilities has exploded over the past seven years. On 9/11, there were two major operational systems: the FBI's Integrated Automatic Fingerprint Identification System (IAFIS) and the Immigration and Naturalization Service's Automated Biometric Identification System, called IDENT, accompanied by a few smaller-scale projects and pilot studies. Today, biometric systems are being used by numerous programs to establish, authenticate and verify identity. The sections below highlight some of the federal government's major operational efforts. While these activities are categorized in this report by agency, most of these efforts required significant interagency collaboration in both their development and operations.

Biometrics Operations Timeline

1993 — Immigration and Naturalization Service (INS) initiates INSPASS using hand geometry at ports of entry to facilitate the inspection of business travelers to the US (using special kiosks and lanes that bypass the normal inspection lanes)

1994 — FBI plans development of the Integrated Automated Fingerprint Identification System (IAFIS).

— INS' IDENT system becomes operational.

1995 — INS uses facial recognition and voice recognition to verify the identity of pre-enrolled persons in vehicles crossing the border at Otay Mesa, California using a special, dedicated lane.

1996 — INS opens the first fully automated Port of Entry at Scobey, Montana relying upon voice verification technology

1998 — DOS begins collecting biometrics from Mexican nationals applying to enter the United States with a Border Crossing Card (BCC).

1999 — FBI's IAFIS major components become operational.

2000 — DoD establishes its Biometrics Management Office (BMO) and Biometrics Fusion Center (BFC).

2003 — DOS begins collecting biometrics from visa applicants through the BioVisa program.

2004 — DHS's US-VISIT program begins collecting biometrics from international visitors at all international air, sea, and land border ports of entry.

— AirNexus kickoff – facilitated travel program operated jointly with the Government of Canada using iris verification at kiosks for pre-enrolled travelers

— DoD's IAFIS-compatible database, Automated Biometric Identification System (ABIS), becomes operational.

— The first statewide automated palm print databases in the United States are deployed in California, Connecticut, and Rhode Island.

2006 — DoD reorganizes the BMO and the BFC into the Biometrics Task Force (BTF).

2007 — TWIC enrollment and issuance begins.

— DOS begins deploying 10-fingerprint collection at all visa-issuing posts.

— US-VISIT begins deploying 10-fingerprint collection at all U.S. airports.

2008 — DHS begins accepting applications for the Global Entry expedited trusted traveler program.

1. Department of Defense (DoD) (Law Enforcement/Intelligence/Access Control)

Today, the DoD recognizes and supports the critical role biometrics plays in national security. Because biometrics makes a difference in the current fight against terrorism, protects the warfighter and the homeland through data sharing with other agencies, and is a tool to bring more effective and efficient business processes to the federal government, it will continue to play a key role in the future security and development of our country and our world.

a. History of DoD Biometrics—Formalizing, Centralizing, Funding, Access Control

The DoD began implementing biometric technologies in 2000 following a feasibility study commissioned in 1999 by the U.S. Congress. This study demonstrated that biometric technologies were an emerging capability that would have a significant impact on the DoD and needed to be formalized, centralized, and funded.

The **Biometrics Management Office** (BMO) was established within the chain of command of the Army's Chief Information Officer (CIO/G-6). The Secretary of the Army was named as Executive Agent (EA) for the DoD, making the BMO the focal point for biometrics for all of the military branches and DoD agencies. The mission at that time focused on Information Assurance (IA), particularly network access.

In the fall of 2000, the **Biometrics Fusion Center (BFC)** opened in Clarksburg, W.Va. Reporting to the BMO, the BFC was tasked with testing commercial biometric products for accuracy and compatibility with DoD information systems. Over the next three years, the BMO and BFC were heavily involved in running pilot projects to evaluate the practicality of using biometric technologies for managing both network and physical access. Work began on designing the backbone architecture needed to pass biometric data securely and quickly between DoD installations and vessels. Development of standards for biometric templates, files, software, and hardware began in earnest. The BMO became a significant contributor in the development of DoD, federal, and international standards for biometrics. At the same time, the BMO began to identify and develop formal policies regarding biometrics as the need arose across the military branches and DoD agencies.

b. DoD Biometrics Post 9/11—Identifying Terrorists, Storing, Analyzing Biometric Data

After the 9/11 terrorist attacks, the DoD developed a vision for using biometrics to lock down the identity of known or suspected terrorists. This represented an expansion of the BMO and BFC mission beyond simply keeping American facilities and networks secure. To accomplish this, the DoD saw the need for a biometric collection and storage system compatible with the FBI's IAFIS. In 2004, the DoD Automated Biometric Identification System (ABIS) became operational. This database gave the DoD a centralized storage point for biometric data collected by the military. In 2006, the BMO and BFC were moved from reporting to the Army CIO to reporting to the Army Chief of Operations (G-3/5/7). At that time, the BMO and BFC were reorganized into the Biometrics Task Force (BTF).

As the ABIS developed, biometric systems that had already been in use for small-scale verification applications were adapted so that the data captured by them would be compatible with ABIS. Operations Enduring Freedom and Iraqi Freedom made it clear that warfighters needed more advanced tools for distinguishing known terrorists and insurgents from friendly populations and that biometric technologies could help fill that need.

The value of the ABIS and various biometric collection and verification platforms has been repeatedly demonstrated since 2004. With data in the ABIS expanding to more than 1.5 million records by summer 2007, biometric matches gave the warfighter a tool to aid in distinguishing between friend and foe. For example, some Iraqi personnel applying for selection to the Iraqi Police Academy were found to have biometric records as terrorists or insurgents. Some detainees in theater were matched to felony records in the United States.

Operational Success: DOD and FBI Partnership

Joint efforts between the DoD and the FBI to compare biometric datasets showed a previously undiscovered trend: numerous individuals that the DoD captures in war zones in Iraq and Afghanistan have prior criminal histories in the United States. This discovery led to even greater collaboration between the two agencies in the war theaters.

"To date, (the Bureau) has developed more than 2,500 latent fingerprints from items such as cordless telephone circuit boards and remote devices -- even batteries and electrical tape. They have made 60 fingerprint identifications and more than 1,000 forensic matches between IEDs."

--FBI Director Robert Mueller, March 28, 2007

As the collection, transmission, and storage systems have matured, the frequency of such matches has increased. At the same time, the response time to answer "Should I detain or not?" has decreased, helping the warfighter to protect himself and other Coalition forces by quickly separating suspected enemies from the general population.

Although the ABIS can quickly determine if there is a biometric match, it cannot determine the value of that match. Is it a match between the fingerprints of a known terrorist and a police academy applicant or between a previously cleared U.S. facility employee and a police academy applicant? The need for this "so what" information has led to new relationships between DoD law enforcement

and the DoD intelligence community. These groups can determine the value of a match and then help get that answer back to the warfighter who needs to know "Should I detain or not?" Much of the work at the BTF focuses on facilitating the architecture, policy, and relationships that get this information quickly back to the warfighter.

Biometric technologies other than fingerprint technologies are also in use throughout DoD. Prototype iris matching has been performed, resulting in unexpected matches when there were no previous connections between the individual records using only fingerprint data. When biometric fingerprint records were examined by certified fingerprint examiners, different reported identities were shown to be the same person. This demonstrates the value not only of iris biometrics, but also that of modality fusion as well.

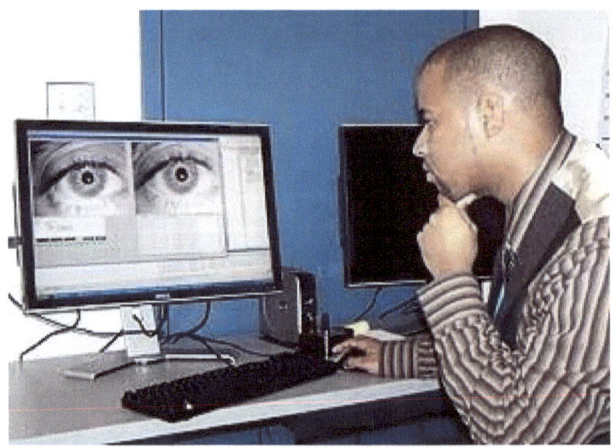

Interagency matches of iris records between the FBI and DoD foreign detainee databases have also yielded results. Agreements between the DOJ and DoD have made this type of data sharing possible. The DoD is moving ahead with establishing not only common technical architectures with non-DoD federal agencies, but also the policies to allow sharing of biometric data while ensuring that legal and privacy rules are followed.

The need for biometric technologies in DoD is clear. As a result, the scope of their deployment will only increase. Along with using biometrics for identifying the enemy, biometrics will soon be used for managing base, building, and network access in accordance with federal guidelines that ensure commonality across the government. Identification of the enemy will also increasingly be a shared governmental function, requiring a common architecture and shared infrastructure across the government. American citizens will not tolerate a situation in which DHS, after taking biometric data, would grant entry into the United States to a person that the DoD can identify as an enemy based on his or her biometric file. A great amount of work is required to tie together biometric and biographic watch lists and the technical architecture to collect and match biometrics across federal agencies. DoD, DHS, and DOJ are meeting these challenges.

c. **Today's DoD Biometrics Structure—Operating and Synchronizing Technologies and Capabilities**

Today, DoD has an integrated structure to program, develop, and synchronize biometric technologies and capabilities and to operate and maintain DoD's authoritative biometric database to support the National Security Strategy.

In October 2006, DoD designated the Director, Defense Research and Engineering (DDR&E), under the Under Secretary of Defense for Acquisition, Technology, and Logistics (USD(AT&L)), as the Principle Staff Assistant for Defense Biometrics with overall responsibility for DoD biometric programs, initiatives, and technologies. The Biometrics DoD Directive of February 2008 further refined these responsibilities and authorities, including appointing the DDR&E lead for interagency coordination, and outlined the policy roles and responsibilities for all DoD biometric stakeholders. The BTF executes day-to-day biometric functions and leads coordination for strategic movement forward for all parts of the DoD. The BTF is supported by multi-Service governance structures that capture Service and user requirements, provide coordination of science and technology efforts, and identify and resolve biometrics-related issues.

2. Department of Homeland Security

> *"On 9/11, America was attacked from within, by 19 men who entered our country, hid among us, and then killed thousands. To stop this from happening again we've taken important steps to prevent dangerous people from entering America. We made our borders more secure, and deployed new technologies for screening people entering America."*
>
> *Remarks by President Bush on the 5th anniversary of DHS.*

The establishment of DHS, a direct response to the 9/11 attacks, was the most sweeping reorganization of the federal government since the start of the Cold War—merging 22 different government organizations into a single department with a clear mission: to protect America from future attacks. DHS is charged with keeping terrorists and their weapons out of the United States while at the same time providing a welcoming environment for the roughly one million international travelers arriving at U.S. ports each day.

A number of organizations within DHS have statutory and regulatory mandates to incorporate biometrics into identity documents for the purpose of freezing identity, searching watchlists, conducting criminal background checks, reducing fraud, improving border and transportation security, and granting benefits and credentialing.

DHS directorates maintain their autonomy and responsibility for planning and managing their biometrics efforts; however, to ensure Department-wide coordination on biometric issues and standards, the DHS Biometrics Coordination Group (BCG) was established. The BCG serves as a focal point for intra-departmental planning and coordination on biometrics RDT&E and deployment to operational end-users. The BCG has been granted delegated authority by the DHS Chief Information Officer and the Under Secretary for Science and Technology to coordinate biometrics technology policy, standards and RDT&E requirements, and to establish a common view of DHS equities on biometrics technology issues before national and international groups.

a. **US-VISIT (Immigration and Border Management/Law Enforcement/ Intelligence)**

CHALLENGE

The United States has more than 300 official ports of entry where nearly a half billion crossings occur every year. The Department of State (DOS) considers more than 9 million visa applications annually. DHS processes nearly 50,000 requests for asylum annually and processes approximately 30,000 applications for immigration benefits every day. The U.S. economy depends on the quick and efficient movement of people and goods across our borders. Among the equally imperative needs for security, law enforcement, travel, immigration, and trade, the United States faces an exceptionally complex challenge that requires unprecedented levels of coordination, advanced technology, innovative thinking, investment, and collaboration.

As the 21st century approached, DHS needed a better system to collect, store, analyze, and share information about international visitors to assess risk and protect the United States from dangerous people. The Department's initial focus included the development of a biometrics-based entry-exit system for international visitors. This system would enhance the security of the United States and ensure the integrity of the immigration and border management system while facilitating legitimate travel and trade and protecting the privacy of visitors to the United States.

SOLUTION

DHS' US-VISIT Program provides biometric identification and analysis services to agencies throughout the immigration and border management, law enforcement, and intelligence communities to accurately identify people and assess whether or not they pose a risk to the United States. Biometrics form the foundation of US-VISIT's services because they are unique, reliable, convenient, and virtually impossible to forge.

The most visible US-VISIT service is the collection and analysis of biometrics—digital fingerprints and a photograph—from international visitors at U.S. visa-issuing posts (collection of the biometrics is handled by the DOS BioVisa Program, and US-VISIT provides the analysis of the data against IDENT) and ports of entry. This service provides US-VISIT customers with the information they need to make efficient and well informed decisions. US-VISIT systems establish and verify international visitors' identities for U.S. Customs and Border Protection (CBP) or DOS consular officers to help them make admission or visa-issuance decisions. U.S. Citizenship and Immigration Services (CIS) uses US-VISIT services to help facilitate requests for immigration benefits. U.S. Immigration and Customs Enforcement (ICE) officers receive credible leads on immigration violators through US-VISIT. US-VISIT establishes and verifies the identities of illegal migrants apprehended by the U.S. Border Patrol along U.S. land borders and the U.S. Coast Guard at sea.

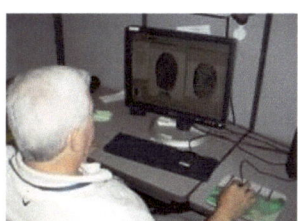
US-VISIT analyzes biometric information collected from locations where terrorists have been, like safe houses or training camps for DoD and the intelligence community, to help them identify terrorists and terror suspects. US-VISIT's Biometric Support Center analyzes fingerprints for federal, state, and local agencies to solve crimes, identify John or Jane Does, and support terrorist investigations.

PROGRAM OVERVIEW

Between 1996 and 2004, Congress passed a series of laws that would enhance the security of the United States and ensure the integrity of the immigration and border management system. DHS established the US-VISIT program in 2003 as a means to better integrate existing information on international visitors and provide decision makers throughout immigration and border management, law enforcement, and intelligence agencies with the right information at the right time and in the right way. Through the collection, storage, analysis, and sharing of biometric-based information, US-VISIT is meeting congressional mandates and the four core goals set when the program began. These goals are to enhance the security of the United States, facilitate legitimate travel and trade, ensure the integrity of the immigration system, and protect the privacy of visitors.

US-VISIT's initial focus was clear and urgent: deploy an electronic, automated, integrated entry and exit capability at U.S. airports and seaports by the congressionally mandated deadline of December 31, 2003. In accordance with this congressionally mandated deadline, US-VISIT deployed the biometric screening capability to U.S. airports and seaports by December 31, 2003, and began collecting visitors' fingerprints and a digital photograph on January 5, 2004.[3] At the same time, US-VISIT began testing biometric exit procedures at airports and seaports. The following timeline demonstrates US-VISIT's incremental deployment of biometrics-based capabilities as of June 2008.

Operational Success Story — Fraud Detection

In March of 2008, a man arrived at New York's JFK International Airport and presented a valid passport and an unexpired visa to the CBP officer. The name on his travel documents did not raise any concerns. However, when his fingerprints were checked through US-VISIT, they did not match the fingerprints associated with the visa he presented. In fact, further investigation showed that he was trying to use the visa belonging to his twin brother, who had no history of criminal or immigration violations. By matching his biometrics, CBP officers learned that this man had been apprehended for taking photos of a U.S. military base, had overstayed the term of his admission on a previous visit to the United States, and had been asked to leave voluntarily following a previous deportation hearing. CBP officers prevented this immigration violator from using fraudulent travel documents to enter the United States and removed him from the country.

2004: US-VISIT deployed biometric entry procedures at all international air, sea, and land border ports of entry.

---

[3] US-VISIT began processing international visitors through the new biometric entry procedures on January 5, 2004, not December 31, 2003, in order to accommodate the high volume of travel through airports and seaports during the holidays.

BioVisa program began.

Biometric entry procedures expanded to include visitors traveling under the Visa Waiver Program (VWP).

2004–2006: US-VISIT tested biometric exit procedures at 14 major airports and seaports.

US-VISIT worked with countries participating in the VWP to meet the congressionally mandated deadline requiring them to issue e-Passports.

2005–2006: US-VISIT tested radio frequency identification (RFID) technology at five land border ports of entry.

2006: US-VISIT deployed e-Passport readers at the necessary airports to process visitors traveling under the VWP.

US-VISIT and the U.S. Coast Guard began testing the use of mobile biometric technology to identify illegal migrants at sea.

US-VISIT, DOJ, and state and local law enforcement agencies began testing the first phase of an effort to make US-VISIT's IDENT and the FBI's IAFIS fingerprint databases interoperable.

2007: US-VISIT tested 10-fingerprint collection at U.S. airports.

US-VISIT's biometric technology enables officers to quickly and efficiently verify that international visitors are who they say they are and do not pose a threat to the United States.

2008:   US-VISIT signed a Memorandum of Understanding with the United Kingdom to collect and transmit biometric and biographic data from U.K. visa applicants at USCIS's Applications Support Centers on behalf of the U.K. government.

US-VISIT published a Notice of Proposed Rulemaking (NPRM) that would establish biometric exit procedures at U.S. airports and seaports.

US-VISIT continues to build on these existing capabilities to continue supporting DHS' mission to protect the United States from dangerous people. By the end of 2008, US-VISIT plans to have the 10-fingerprint collection capability at all U.S. ports of entry, begin the second phase of IDENT/IAFIS interoperability, and provide the DHS Secretary with a report on the challenges and opportunities for deploying biometric exit procedures at land border ports of entry. Additionally, DHS plans to begin deploying biometric exit at U.S. airports in 2009. US-VISIT will continue to work with other countries as they develop biometrics-based immigration and border management systems to learn from and share US-VISIT's best practices.

PRIVACY POLICY

DHS protects the biometric and biographic information provided by travelers and ensures that their privacy is protected in a manner consistent with all applicable privacy laws and regulations. Personal information is kept secure and confidential, and appropriate security controls ensure that the data are not used or accessed improperly.

US-VISIT publishes Privacy Impact Assessments (PIAs) to ensure that personally identifiable information (PII) is used appropriately, protected from misuse and improper disclosure, and destroyed when no longer needed. PIAs are updated as needed to ensure they remain current with any changes to the programs and systems.

US-VISIT's dedicated privacy officer is responsible for the program's compliance with privacy laws and procedures, as well as creating a culture within the program where privacy is inherently valued, treated as a fundamental right and obligation, and embedded into planning and development processes. Information on the US-VISIT privacy program is available at *www.dhs.gov/us-visit*.

US-VISIT complies with all environmental laws and regulations. Environmental Impact Assessments conducted prior to deployment of every phase of the program have found no adverse impacts.

AUTHORIZING LEGISLATION AND FUNDING

US-VISIT received $330 million in FY04, $340 million in FY05, and $336 million in FY06, and Congress appropriated $362 million for FY07 and $475 million for FY08. The following laws are relevant to the mission and goals of US-VISIT.

- The Illegal Immigration Reform and Immigrant Responsibility Act of 1996 (IIRIRA), Public Law 104-208

- The Immigration and Naturalization Service Data Management Improvement Act of 2000 (DMIA), Public Law 106-215

- The Visa Waiver Permanent Program Act of 2000 (VWPPA), Public Law 106-396

- The Uniting and Strengthening America by Providing Appropriate Tools Required to Intercept and Obstruct Terrorism (USA PATRIOT) Act, Public Law 107-56

- The Enhanced Border Security and Visa Entry Reform Act of 2002 (EBSVERA), Public Law 107-173

- The Intelligence Reform and Terrorism Prevention Act of 2004 (IRTPA), Public Law 108-458, Section 7208

b. **Transportation Worker Identification Credential (TWIC) (Access Control)**

CHALLENGE

Since 9/11, DHS has focused time and attention on enhancing the security of U.S. ports, particularly because of the role the ports play in the U.S. economy. Each day, $1.3 billion worth of goods move in and out of U.S. ports. In addition, many major urban centers (more than half of the U.S. population) and significant critical infrastructure are in proximity to U.S. ports or are accessible by waterways. As points of the entry and exit program, they are critical nodes that affect terrorist travel and transiting of material support or weapons. The economic, physical, and psychological damage that would result from a significant terrorist attack targeting maritime commerce or exploiting America's vulnerability to sea strikes is difficult to estimate, but the stakes are high. A significant breakdown in the maritime transport system would send shockwaves throughout the world economy.

Maritime security requires a partnership between DHS and all other parties involved in the operation of U.S. ports, including transportation employees. DHS needed to create a program ensuring those with unrestricted access to the nation's ports did not pose a threat to national security.

SOLUTION

TWIC is a vital security measure that will help ensure that properly vetted individuals are allowed access to the nation's transportation infrastructure, while denying this access to individuals who pose a threat, do not require unescorted access, or do not warrant access to secure areas of the nation's maritime transportation system.

TWIC was established by Congress in 2002 through the Maritime Transportation Security Act (MTSA) and is administered by the Transportation Security Administration (TSA) and the U.S. Coast Guard. The TWIC credentials are tamper-resistant biometric cards that will be issued to workers who require unescorted access to secure areas of ports, vessels, outer continental shelf facilities, and all credentialed merchant mariners. It is anticipated that more than one million workers (including longshoremen, truckers, port employees, and others) will be required to obtain a TWIC.

The TWIC contains two biometric templates of a person's fingerprint. These templates are stored on the card in a format that is enciphered using a card-specific TWIC privacy key. To confirm a cardholder's identity and ensure it matches the stored biometrics, the data on the card are retrieved, deciphered, verified, and matched against a live finger.

TWIC uses biometrics for two primary identification purposes: background screening and verification. Background screening occurs prior to the issuance of a TWIC and encompasses an FBI criminal history records check and a check of DHS' IDENT database. Post-issuance, biometrics may be used at access control points to ensure that the biometrics of the individual attempting to use the TWIC match those stored within the credential.

PROGRAM OVERVIEW

The TWIC final rule was issued in 2007 by TSA. Enrollment and issuance began at the Port of Wilmington, Del., on October 16, 2007, and will continue through calendar year 2008 and part of 2009.

The TWIC rule involves the following:

- TSA will collect a worker's biographic and biometric information, including: 10 fingerprints, name, date of birth, address, phone number, photo, employer, job title, and, if the worker is not a U.S. citizen, other appropriate information to be able to authenticate the worker's immigration and work authorization status.
- All individuals that require unescorted access to secure areas of port facilities and vessels regulated under the Maritime Transportation Security Act are required to have a TWIC. This includes longshoremen, port operator employees, truck drivers, and rail workers. U.S. merchant mariners who hold an active Merchant Mariner's Document, Merchant Mariner's License, Certificate of Registry, Standards of Training, or a Certification and Watchkeeping Endorsement are also required to obtain a TWIC.
- Background checks are performed and include a review of criminal history records, terrorist watch lists, legal immigration status, and outstanding wants and warrants.
- TWIC uses smart card technology and includes a worker's photo, name, biometric information, and multiple fraud protection measures. The card's technical specifications are consistent with most Federal Information Processing Standards Publication 201-1 requirements and will be interoperable with other federal credentials built to those standards.
- The program is expected to cover approximately 1.2 million workers and is funded through user fees. The fee for TWIC will be $132.50, and it is valid

for five years. Workers with current and comparable background checks will pay a reduced fee of $105.25.

- Port facility and vessel owners and operators are required to implement TWIC into their existing access control systems and operations, purchase and use card readers, and update their approved security plans.

The TWIC program is progressing steadily and has opened more than 130 fixed enrollment centers and dozens of mobile sites nationwide. To date, more than 350,000 workers have enrolled. Thousands more are processed each week.

The U.S. Coast Guard issued the Merchant Mariner Credential (MMC) rule on the same day as the TWIC final rule. The MMC regulation works in conjunction with TWIC to streamline the current credentialing process for merchant mariners, as all U.S. merchant mariners will be required to obtain a TWIC. The TWIC will meet/support the identity verification requirements for MMC holders. The TWIC will also support electronic verification of MMC attributes in the future. The Coast Guard and TSA are streamlining the process for the two credentials to reduce costs, duplication of effort, and processing time for mariners.

PRIVACY POLICY

Privacy and the security of personal information are critical to the TWIC program. Information collected at the enrollment center or during the pre-enrollment process, including the signed privacy consent form and identity documents, is scanned into the TWIC system for secure storage. Information is encrypted and stored at a secure government facility using methods that protect the information from unauthorized retrieval or use.

TWIC has published a PIA to ensure that personal information is used appropriately, protected from misuse and improper disclosure, and destroyed when no longer needed. The PIA can be found on the TSA website at www.tsa.gov.

AUTHORIZING LEGISLATION

The following laws are relevant to the mission and goals of the TWIC.

- The Maritime Transportation Security Act (MTSA), Public Law 107–295

- Uniting and Strengthening America by Providing Appropriate Tools Required to Intercept and Obstruct Terrorism Act (USA Patriot Act), Public Law 107-56

- The Aviation Transportation Security Act (ATSA), Public Law 107-71

c. Global Entry (Immigration and Border Management)

CHALLENGE

While a rigorous inspection process does increase the safety of the nation, long lines can result in irritated travelers, missed connections, and negative impressions of the United States. This can be especially cumbersome for frequent international travelers who pose no risk to the country. How can we facilitate U.S. citizens who are not a threat?

SOLUTION

Global Entry is a pilot program managed by U.S. Customs and Border Protection (CBP) that allows pre-approved, low-risk travelers expedited clearance upon arrival into the United States. Currently, only U.S. citizens and lawful permanent residents are eligible to join. Upon returning from international travel, Global Entry-enrolled travelers may bypass the regular passport control line and proceed directly to the Global Entry kiosk. The Global Entry process will require participants to present their machine-readable U.S. passport or permanent residency card, submit their fingerprints for biometric verification, and make a customs declaration at the kiosk's touch screen. The kiosk will compare the fingerprints presented to the fingerprints on file to confirm the traveler's identity.

Upon successful completion of the Global Entry process at the kiosk, the traveler will be issued a transaction receipt and directed to baggage claim and the exit unless chosen for a selective or random secondary referral.

On June 10, 2008, Global Entry operations began at John F. Kennedy International Airport, Washington Dulles International Airport, and George Bush Houston Intercontinental Airport.

PRIVACY POLICY

The information collected through the online application is secured in the Global Online Enrollment System (GOES) as the system of record for CBP trusted traveler programs. The personal information provided by applicants, including the fingerprint biometrics, may be shared on a need-to-know basis with other government and law enforcement agencies in accordance with applicable laws and regulations. The personal information collected through GOES is maintained in a Privacy Act system of records that was last published in the Federal Register on April 21, 2006, (71 FR 20708). CBP has also published two PIAs that cover this pilot on the DHS Privacy Office website at http://www.dhs.gov/xinfoshare/publications/editorial_0511.shtm. In addition, an update addressing online functionality of the enrollment process was posted to the DHS Privacy Office website on November 1, 2006. The applicant's biometrics are stored in DHS' Automated Biometric Identification System (IDENT). The IDENT Privacy Act System of Records Notices (SORNs) was last published on June 5, 2007, (72 FR 31090).

AUTHORIZING LEGISLATION

The following laws are relevant to the mission and goals of Global Entry.

- Consolidated Appropriations Act, Public Law 110-161
- Paperwork Reduction Act, 44 U.S.C. 3501 *et seq.*
- Federal Information Security Management Act, Public Law 107-347
- Service Data Management Improvement, Public Law 106-205

3. Department of Justice (Law Enforcement and Intelligence)

In September 2001 the DOJ, through the FBI Criminal Justice Information Services (CJIS) Division, maintained the largest and most advanced biometric database in the world, containing fingerprint biometrics linked to criminal history records contributed by all 50 states and US territories. CJIS provided biometric identification services and criminal history information services primarily to law enforcement, but also to civil customers for employment and licensing purposes.

In the months following 9/11, Congress tasked the Attorney General with leading a number of interagency studies and making recommendations on the feasibility of expanding the use of biometric identification services for additional civil purposes such as visa screening. Today, the DOJ, DOS, DHS, and DoD all have established biometrics programs to support identification and screening requirements. The FBI CJIS Division is making a major investment in "Next Generation Identification," to expand identification services and investigation services for law enforcement and counter terrorism purposes, and to keep pace with the growing demand for biometric services for civil purposes.

CHALLENGE

Every day, local, state, tribal, and federal law enforcement agencies in the United States arrest more than 50,000 people. There is a limited amount of time to identify and link them to any outstanding warrants or criminal history. Additionally, well over 60,000 people a day apply for positions of trust, visas to visit the United States, for citizenship, etc. In each case, a check has to be made to determine if there are any facts that would make them unsuitable. The FBI meets these identification challenges through electronic processing of fingerprint-based background checks by its CJIS Division using the IAFIS.

The events leading up to 9/11 showed that these databases and searches were neither comprehensive enough nor rapid enough to support all counterterrorism challenges. Files have to be exchanged with DHS, DOS, and others to ensure that checks made by one department would not miss known or suspected terrorists (KSTs), persons with criminal backgrounds, etc. Biometric-based information also needed to be better coordinated among DoD and the intelligence community in order to "connect the dots."

Other U.S. government agencies needed to access the rich database maintained by CJIS, but were impeded by several obstacles, including response time and the then minimum number of 10-fingerprint images needed for a background check. In many instances where an on-the-fly background check was required (e.g., at a border crossing), CJIS' two-hour response time was not acceptable. While CJIS was achieving substantially faster results than their advertised maximum turn-around time for criminal transactions, they were not in the one- to two-minute range required for transactions at border crossings or in the field where DoD warfighters encounter unknown but suspicious persons.

Not every U.S. government agency had the equipment, facility, or need to capture all 10-fingerprint images for background checks. The warfighter or border agent does not always have the time and equipment a booking officer in a police station might have. Thus, for many years, DoD and DHS (and its predecessors) captured transactions with just two fingerprints, and IAFIS could not process these transactions. The mandatory format for collection and submittal to IAFIS was based on the inked fingerprint card format of all 10 fingers being rolled as well as the same fingers being collected as plain impressions (AKA "sequence slaps"). The following graphic shows the paper format. There is a corresponding electronic format developed in concert with NIST. IAFIS now accepts and processes digitally-collected rolled and flat fingerprint submissions.

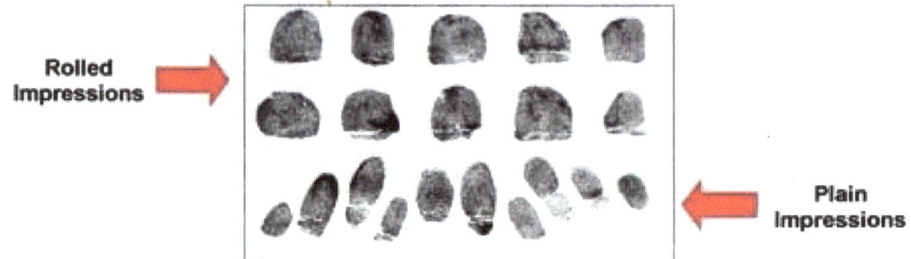

The CJIS Division's identification system, the program, IAFIS, is nearly 10 years old, and the algorithms are even older. Newer algorithms would lead to better matching performance. IAFIS is able to process only fingerprints, yet the government is moving toward a multi-biometric collection protocol that includes fingers, palms, faces, and irises. The challenge therefore was two-fold: to modify existing IAFIS processing to allow government partners to rapidly search millions of files with less than 10 rolled fingerprints and to move to a newer platform that will improve matcher accuracy and use other biometrics in addition to fingerprints.

SOLUTION

With 9/11, the emphasis of the FBI's mission was refocused to make the national security mission as important as the criminal investigation mission. CJIS' mandate expanded from the identification of criminals to include the identification of KSTs and other individuals whose primary goal was the destruction of the freedoms and privileges that are the backbone of America. The mandate also included enhanced data sharing requirements as outlined in the USA PATRIOT Act.

Operational Success Story – Outstanding Warrant

A man applied for asylum at a U.S. Asylum Office. When the man's fingerprints were run against US VISIT's database, which was linked to the IAFIS database, they revealed that this man had an extensive criminal record, including charges for rape, assault and an outstanding warrant in Maryland for kidnapping. Although he had used three aliases and a different date of birth to try and evade detection, his fingerprints confirmed he was the man wanted for kidnapping. As a result of US VISIT's positive identification of this person, the asylum office contacted Immigration and Customs Enforcement (ICE), and ICE arrested the man.

The use of biometrics became even more focused when HSPD-12 was issued on August 27, 2004. As with many other programs, the issuance of a PIV would be dependent on an applicant passing a fingerprint-based background check on the CJIS records.

The USA PATRIOT Act directed the Attorney General to commission a study on the feasibility of using biometric identifiers to identify people as they attempt to enter the United States, which would be connected to the FBI's database to flag suspected criminals. Another study was commissioned to determine the feasibility of providing airlines with names of suspected terrorists before they boarded flights. This created a requirement for greater data sharing between CJIS, DoD, DOS, and DHS.

In response, CJIS and DHS are working together to make the IAFIS fingerprint database and the US-VISIT's IDENT database interoperable. In 2007, the US-VISIT Program began to test 10-fingerprint collection using identification-slaps. The IAFIS was modified to accept the submission of these identification-slap fingerprint images for applicant background checks. These are easier and faster to collect than the traditional rolled impressions. The number of these submittals has grown to approximately 48,000 per day; 30,000 per day from DOS and 18,000 per day from DHS. This number is expected to rise to almost 78,000 per day by the end of 2008.

The CJIS Division has developed a 22-pound *Quick Capture Platform* for contemporaneous biometric collection and search from the field by FBI Hostage Rescue Teams. These platforms have collection devices for multiple modalities and satellite links for remote searches of the IAFIS. Currently, 49 of these units are operationally deployed with approximately 20 of them deployed in Afghanistan alone.

In February 2008, the CJIS *Flyaway Team* deployed to Afghanistan for a 90-day mission to obtain data from the Afghan National Police and the Afghan National Army. These biometrics will be the baseline for the Automated Fingerprint Identification System to be established for the government of Afghanistan through the DoD and the FBI. CJIS staff have visited 29 countries this year through the Foreign Fingerprint Exchange Program.

The IAFIS now has unique databases for agency partners, which allows them to use the resources of the extensive CJIS Division databases. For example, in 2004 the DoD ABIS became operational and is compatible with the IAFIS and co-housed with the IAFIS in Clarksburg, W.Va. CJIS is supporting the global war on terrorism by working with DoD in searching latent fingerprints from improvised explosive devices (IEDs) to identify persons involved in anti-Coalition force activities.

The FBI maintains a website at http://www.fbibiospecs.org/fbibiometric/biospecs.html dedicated to providing the most up-to-date information regarding FBI biometric standards initiatives from the CJIS Division, Technology Evaluation Standards Test Unit. Current offerings include the following.

- Electronic Biometric Transmission Specification (EBTS)
- The Registry of USG Recommended Biometric Standards
- IAFIS Certified Products List

PROGRAM OVERVIEW

The FBI became the national repository for fingerprints and related criminal history data in 1924 when 810,188 fingerprint records from the National Bureau of Criminal Identification and Leavenworth Penitentiary were consolidated to form the nucleus of the FBI's files. Since then, the FBI's fingerprint files have grown to become the world's largest biometric repository with associated criminal history information. Fingerprint identification services, which had steadily increased over the years, became even more important following 9/11.

The FBI CJIS Division maintains both criminal and civil fingerprint records in separate databases. Today, the master criminal fingerprint file contains the records of approximately 56.4 million individuals, while the civil file represents approximately 20 million fingerprint submissions. The civil file predominantly contains fingerprints of past and present U.S. military personnel and present and former federal government employees.

The paper-based process changed on July 28, 1999, with the CJIS Division implementation of the IAFIS. The IAFIS provides an up-to-date, integrated system to respond to the needs of the local, state, tribal, federal, and international criminal justice and authorized non-criminal justice agencies. It houses the largest collection of digital representations of fingerprint images and associated criminal history information in the world. The current operation supports electronic submission of fingerprint identification data to IAFIS and an electronic response to the inquiring agency. An electronic response is normally sent within two hours of a criminal identification request and within 24 hours of an electronic civil submission.

Originally designed to process 62,500 fingerprint submissions daily, the IAFIS now averages approximately 90,000 fingerprint transactions per day. A record was achieved on July 23, 2008, when 163,089 transactions requesting searches against the criminal repository were completed within a 24-hour period. Each day, approximately 8,000-10,000 new records are added to the criminal repository.

Prior to 9/11, the FBI receipts averaged 15.4 million fingerprint submissions annually. Approximately 7.4 million, or 48%, were civil fingerprint submissions. During FY07, the FBI received a total of 26.1 million fingerprint submissions. Of this total, 56%, or approximately 14.5 million, were civil fingerprint submissions.

The FBI continues to improve existing processing to provide the most reliable and accurate information possible on a system that is now nine years old. To remain responsive to law enforcement and other customer needs, CJIS must embrace the advances in identification technology. It is essential that enhancements be made to the FBI identification program.

The Next Generation Identification (NGI) Program
Advances in technology and the changing business needs of IAFIS customers have highlighted the need for a next generation of identification services. To further advance its biometric identification services, CJIS, along with guidance from its user community, has established the vision for the Next Generation Identification (NGI). The NGI Program will improve the current functionality of the IAFIS and provide new identification modalities to enhance the accuracy and quality of biometric records. NGI will offer state-of-the-art, multimodal biometric identification services through numerous initiatives, including:

Advanced Fingerprint Identification Technology
Advanced Fingerprint Identification Technology will provide faster, more efficient identification processing, increased search accuracy, and improved latent processing services. As a new feature, the Repository for Individuals of Special Concern (RISC) will provide the capability to search two or ten fingerprints against fingerprints of wanted persons, KSTs, and sex offender registry subjects with a response returned in seconds. This service will be expanded to allow for the same type of rapid search against other special populations such as persons of national security interest.

Evaluating the Effectiveness of Multimodal Biometrics
The NGI Program will advance the integration strategies and indexing of additional biometric data, providing a framework for a future multimodal identification system. Collection and use of additional biometrics must be cost-effective and demonstrably enhance the accuracy and quality of fingerprint records. This will permit identifications based on not just fingerprints, but faces, irises, palm prints, and the fusion of these identification techniques.

The Biometric Center of Excellence
The FBI'S Science and Technology Branch has established a Biometric Center of Excellence (BCOE) within CJIS to support U.S. government-wide biometric and identity management activities. The BCOE will facilitate research, development, and training activities that relate to biometric technologies. Plans for the BCOE include the

construction of the Biometric Technology Center in Clarksburg, W.Va. The Center will house the FBI biometric operations and the DoD Biometric Task Force with the potential addition of new partner agencies.

Interim Data sharing Model (iDSM)

> **Operational Success Story – IDENT/IAFIS Interoperability**
> In Boston, officers arrested a subject for breaking and entering. Through the one-step biometric submission process that the pilot provides, officers in Boston and ICE's LESC were automatically notified of the subject's extensive history of criminal and immigration violations. Though the subject had used seven different aliases
> and nine dates of birth over the years to evade detection, the subject's biometrics revealed an extensive criminal record and two previous deportations. The subject is now detained.

To take advantage of the benefits of interoperability, CJIS partnered with several agencies to implement the iDSM. The iDSM includes the sharing of the following datasets among appropriate agencies and departments.

- IAFIS Wants and Warrant records (DOJ)
- KST records (DOJ) – added 8/09/2007
- IDENT Expedited Removal records (DHS)
- Category I Visa Critical Refusals records (DOS)

Several local agencies as well as federal agencies are participating in a pilot program for iDSM searches. These agencies include the following.

- Boston Police Department online 09/03/2006
- Dallas County Sheriff's Office online 11/01/2006
- Office of Personnel Management online 12/01/2006
- Harris County Sheriff's Office online 02/01/2007
- DoD online 04/03/07

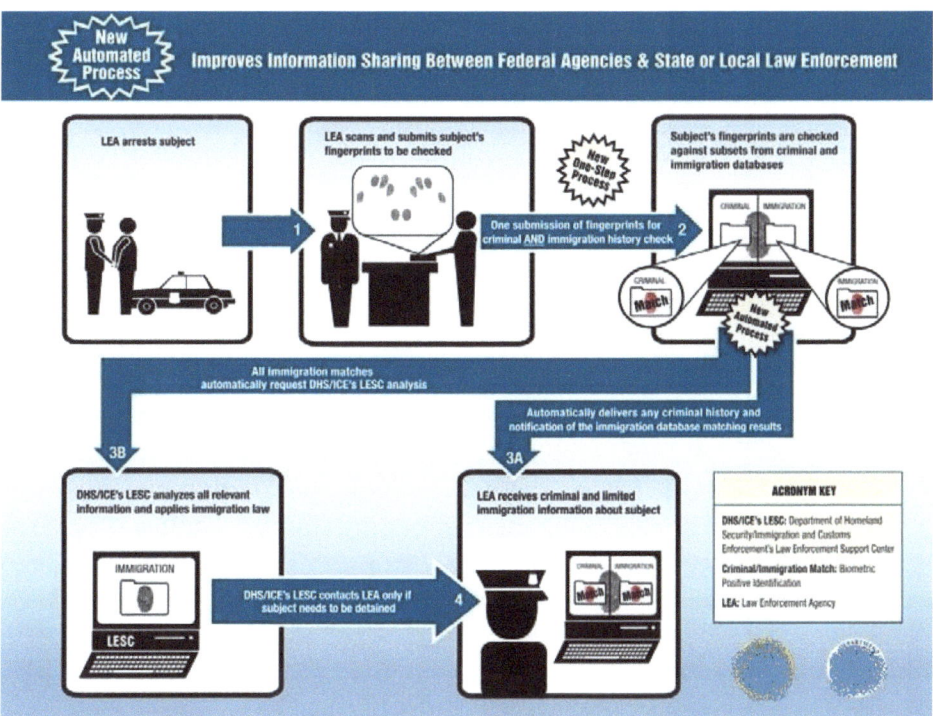

Immigration Violator File

As of April 2008, all incoming IAFIS submissions are being searched against the NCIC Immigration Violator File (IVF) records that have associated fingerprints. The IVF contains name-based records of subjects that are a high enforcement priority for the DHS' Immigration and Customs Enforcement.

PRIVACY POLICY

The FBI protects the privacy of individuals whose biometric templates are in IAFIS according to the Privacy Act of 1974, tempered by the need to share information between different government agencies as specified in the Intelligence Reform and Terrorism Prevention Act 2004.

SUMMARY

The acceptance and success of biometrics have increased law enforcement enrollments and searches both nationally and internationally. Responding to the challenges of terrorists and transnational criminals has expanded both domestic and international data sharing needs. The FBI has responded to these needs with a refocusing of priorities, enhancements of business practices and the related technology, and more cooperative sharing measures. As national boundaries blur, the political, technical, and legal frameworks become more challenging. The CJIS Division continues to meet these challenges. In the next few years, the NGI system will revolutionize the level and scope of services that CJIS provides.

4. Department of State (Immigration and Border Management and Intelligence)

Similar to previously discussed federal departments, the DOS also had small-scale biometrics efforts prior to 9/11, but activities have greatly increased since that time. Examples of operational efforts are described below.

a. Biometric Visa Program

The Biometric Visa (BioVisa) Program was developed and implemented to enhance the security of the U.S. visa and thereby strengthen the border security of the United States. The legislative bases for the BioVisa Program are section 403(c) of the USA PATRIOT Act, which mandated biometric screening for visa applicants, and section 303(a) of the Enhanced Border Security and Visa Entry Reform Act, which required the Secretary of State to use biometric identifiers for all visas issued to aliens.

b. Border Crossing Card Program

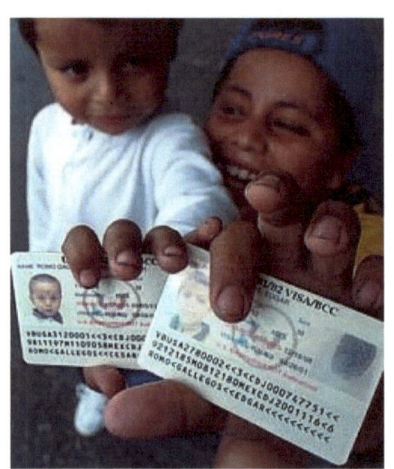

Since 1998, DOS has had experience using biometrics at consular posts in Mexico in the visa process through the Border Crossing Card (BCC) Program, which was mandated by the Illegal Immigration Reform and Immigrant Responsibility Act. The BCC, issued only to Mexican nationals, serves as a visa. Under the BCC Program, two index fingerprints and a photo were captured of BCC applicants and cleared against the fingerprint watch list of DHS' IDENT, which contains the fingerprints of suspected terrorists, wanted persons, and immigration law violators. IDENT also stores the fingerprints of BCC applicants. The BCC Program served as the model for the BioVisa Program.

c. BioVisa and US-VISIT as Partner Programs

The BioVisa Program was established as a partner program with DHS' US-VISIT program. In rolling out the BioVisa Program, DOS started deployment of equipment and capturing two index fingerprints of visa applicants in September 2003. Thirteen months later, on October 7, 2004, all posts issuing nonimmigrant and immigrant visas were capturing fingerprints of applicants. From the very beginning, the BioVisa Program has been responsible, through the results of the fingerprint checks against the IDENT watch list, for the refusal of visas to many thousands of ineligible applicants who would likely have succeeded in obtaining visas had it not been for the fingerprint checks. For example, in the first six months of 2008, there were 12,932 matches of visa applicant fingerprints with fingerprints in IDENT, and in almost all of those cases the visas were refused.

Operational Success Story – Multiple Aliases

In 2007, an individual applied for a visa at a U.S. Embassy. When the person's fingerprints were checked against the US VISIT biometric database, it was revealed that the person had been denied a visa just one day earlier under a different name. Therefore, the person was denied the visa due to willful fraud and misrepresentation.

d. Biometric Identity Verification at Ports of Entry

Aside from the screening of visa applicants against the IDENT watch list, the BioVisa Program also enables CBP officers at ports of entry (POEs) to match the fingerprints of a person presenting a visa with the fingerprints in the IDENT database that were captured at the time of visa issuance. This ensures that the person presenting the visa at the POE is the person to whom the visa was issued, thus preventing visa fraud. This biometric identity verification at POEs guarantees the integrity of the U.S. visa because it has essentially eliminated the possibility of visa fraud through counterfeit or photo-substituted visas or through the use of valid visas by imposters.

e. Issued Visa Records Viewed at Ports of Entry

To ensure the security of valid visas issued prior to the beginning of the BioVisa Program, DOS and the US-VISIT Program implemented a separate procedure by which the visa data of all issued visas worldwide, which are replicated within 10 minutes to the Consular Consolidated Database at DOS and relayed directly to the DHS Treasury Enforcement Telecommunications System (TECS), are made available for display to CBP officers at primary inspection. Under this procedure, when the passport or visa is scanned at primary inspection, the visa data, including the photo, are retrieved from TECS and displayed on the CBP officer's screen. This procedure, in addition to the matching of the person's biometrics, prevents the counterfeiting or photo substitution of visas issued prior to the beginning of the BioVisa Program.

f. DOS Facial Recognition System Screens Photos of Visa Applicants

Under the BioVisa Program, diplomats, certain other government officials, children under age 14, and persons age 80 and over are exempt from fingerprints during the visa application process. The photos of all applicants exempt from fingerprinting are cleared against a photo watch list of KSTs in the DOS Facial Recognition System. To combat visa fraud, visa applicant photos are also checked through the Facial Recognition System against over 68 million photos in the Consular Consolidated Database.

g. BioVisa Program Transition from Two to Ten Fingerprints

To build on the success of the BioVisa and US-VISIT programs, federal agencies, working through the Homeland Security Council, decided in 2005 that there should be a transition from two to ten fingerprint collection. Collection of 10 fingerprints enables the following: 1) 10 fingerprints provide additional biometric information that can improve the accuracy of the IDENT system; 2) 10 fingerprints provide additional matching opportunities with latent fingerprints collected from terrorist or other crime scenes; and 3) 10 fingerprints can be checked against the full IAFIS criminal master file to prevent issuance of visas to persons with criminal records.

h. Ten Prints Screened Against KST Latents in IDENT

During 2007, Consular Affairs transitioned all visa-issuing posts from collection of two to ten fingerprints. The 10 fingerprints sent to IDENT are checked against all available KST latent fingerprints, as well as latent prints from federal crime scenes. Moreover, latent fingerprints collected from IEDs in Iraq and Afghanistan are transferred to IDENT to be used in checks against visa applicant fingerprints.

i. BioVisa 10 Prints Advance IDENT-IAFIS Interoperability

In January 2008, the 10 fingerprints collected from visa applicants began to be searched against the FBI's IAFIS criminal master file. The process for this is that the 10 fingerprints continue to be sent across the interface from Consular Affairs to IDENT, which searches them against IDENT but also relays them to IAFIS for a search. The results of the IAFIS search are returned to Consular Affairs through the IDENT interface. In this manner, an additional benefit in IDENT-IAFIS interoperability has been achieved.

j. 10 Print Screening Against IAFIS

Since 2002, the FBI CJIS Division has provided Consular Affairs with names of wanted persons and persons with criminal history records for inclusion in the Consular Lookout and Support System (CLASS), which is the name-based lookout system against which the names of all visa applicants are screened prior to issuance of visas. However, persons with criminal records could avoid detection by the CLASS screening by obtaining passports in different identities. The BioVisa transition to 10 fingerprints and the screening of the 10 prints against IAFIS has prevented criminals from being able to conceal their criminal records when applying for visas. In the first six months of 2008, the fingerprints of more than 4 million visa applicants were screened against IAFIS; 27,912 of those visa applicants had records of arrest and prosecution (RAP sheets). Many of these RAP sheets involved crimes that rendered the visa applicants ineligible for a visa.

k. BioVisa Program Essential for Border Security

By preventing ineligible applicants from obtaining visas and by enabling biometric identity verification of persons presenting visas at ports of entry, the Biometric Visa Program has proven to be an unqualified success in strengthening the border security of the United States.

5. Personal Identity Verification Credential (access control)

Homeland Security Presidential Directive 12 (HSPD-12), signed by the President in August 2004, established the requirements for a common identification standard and credentials to be issued by federal agencies to federal employees and contractors to gain physical access to federal facilities and logical access to systems and networks. The directive specified that the technical requirements for the secure credential meet four control objectives:

- Is issued based on strong criteria for the verification of an individual's identity;
- Is strongly resistant to identity fraud, tampering, counterfeiting, and terrorist exploitation;
- Can be authenticated electronically; and
- Is issued only by providers whose reliability has been established by an official accreditation process.

NIST was directed by the HSPD-12 to create standards and requirements for the security and interoperability of the cards and processes required for the government-wide implementation of HSPD-12. After significant consultation, both within the government and with the private sector, NIST issued Federal Information Processing Standard (FIPS) 201, The Personal Identity Verification Standard, in February 2005. NIST has issued additional technical specifications to ensure that the cards, data stored on the cards, and data interfaces are standardized across government implementations. The General Services Administration (GSA) established the FIPS-201 Evaluation Program in May 2006 to evaluate commercial products and services for conformance to the normative requirements of FIPS-201.

For the first time in history, the President's annual budget request to Congress was transmitted electronically on February 4, 2008. The Executive Clerk used an HSPD-12 approved credential to digitally sign the electronic transmittal of the budget to Congress, thus proving the document's authenticity.

The federal government has established 23 categories of products and services (e.g., smart cards, card readers, fingerprint scanners, face image capture equipment, card printing equipment, etc.) that require evaluation and testing for conformance to FIPS-201 requirements. Commercial industry has responded to the FIPS-201 requirements quickly and effectively. There now are more than three-hundred compliant products approved for government-wide use for the implementation of HSPD-12. The FIPS-201 Approved Products List is available at http://www.idmanagement.gov.

Advancing and Utilizing Privacy Theory

To some individuals, biometric information represents the most worrisome of all forms of personally identifiable information (PII). The Subcommittee, as well as agencies with operational missions, takes this concern quite seriously and has worked to simultaneously advance privacy technology theory and to integrate these new concepts into operational system planning and oversight. Subcommittee members met with the DHS Privacy Office soon after it was established to ask for assistance and guidance on how to better approach this problem. The Privacy Office immediately recognized both that solving the biometrics privacy problem is critical, and that this solution could easily be adapted to less privacy-sensitive technologies. Since that time, privacy work on biometrics has advanced both through the Subcommittee and among agency privacy offices. A highlight of activities is provided below.

> ""Privacy means more than "private" – it is not limited to keeping a secret. Most conceptions of secrecy assert that once the secret is revealed it is available for any public use (the individual "owner" of the secret loses all claims of control over the information). However, privacy claims can cover information and activities involving others (for example, bank accounts held by banks, medications known to doctors and pharmacists, etc.). In the biometric context, privacy protection governs the use of personal information that is shared (not "secret"). In response, the biometrics community must work to implement policies and processes that effectively govern the appropriate use of data, individually and in its aggregate. These policies and procedures should be clearly communicated to all affected constituencies.""
>
> *From The National Biometrics Challenge document of August 2006*

1. Building a Conceptual Foundation

The first step toward advancing biometrics privacy was to establish a common reference. Immediately after 9/11 and for some time thereafter, there existed a state of cross-talk between biometrics and privacy experts. Biometrics experts, mostly scientists, attempted to study and explain privacy issues but were unsuccessful as they didn't truly grasp the legal and social ramifications. Privacy experts, mostly lawyers, attempted to study and explain biometric issues but were unsuccessful as they didn't truly grasp the technology's capabilities and limitations. Neither group of experts spoke in a manner that was understandable by the other.

The subcommittee brought these two groups of experts together so that they could engage in a series of cross-training discussions, which led to an enhanced understanding of basic theories and science, thus enabling advanced application on operational systems. To promote this understanding throughout and beyond the U.S. government, the subcommittee later developed a public paper entitled *Biometrics and Privacy: Building a Conceptual Foundation*, which was released in 2006. The paper attempts to connect privacy and biometrics at a structural level so that both fields can be understood within a common framework. The paper provides a general overview of both privacy and biometrics, and offers a perspective from which to view the convergence of both. The goal is to provide a context in which details and future developments can be placed and better understood.

2. Privacy Impact Assessments

Privacy Impact Assessments (PIAs) are a key aspect of the federal government's privacy compliance efforts. Section 208 of the e-Government Act requires all federal agencies to conduct and complete PIAs for all new or substantially changed technologies that collect, maintain, or disseminate PII. The PIA process forces system owners and developers to ensure that they have consciously incorporated privacy protections throughout the entire system development lifecycle. A PIA provides an analysis of how PII is collected, stored, protected, shared, and managed. For example, the PIA process provided greater transparency into CBP's implementation of the air phase and the land/sea phase of the Western Hemisphere Travel Initiative (WHTI). Privacy Officers coordinate the completion of PIAs for the Department and the components and must approve them prior to systems being implemented and/or adjusted.

As privacy compliance has matured, so too has the content and procedures for conducting a PIA. The lessons learned from previous reviews have translated into better content for each subsequent PIA process.

3. Privacy Technology Implementation Guide

To better inform managers of technology projects on how to approach the PIA process, the DHS Privacy Office developed the Privacy Technology Implementation Guide (PTIG) in 2007. The PTIG incorporates privacy protection considerations, organized according to privacy compliance requirements, and presents those considerations in the context in which technologists will encounter them: in the management and development of operational systems.

This guide does not dictate additional mandates for system development. Instead, the PTIG offers a new method of raising awareness regarding what "privacy protection" means in the context of managing and developing operational systems and, through that awareness, initiating the process of privacy compliance earlier in the system development life cycle and more thoroughly across the overall process of deploying systems.

The goal of this guide is to raise awareness of privacy issues for those working directly with technology and to present additional considerations that, if addressed directly and early in system development, can improve the effectiveness and efficiency of complying with privacy protection requirements.

4. International Privacy Workshops

Biometrics privacy issues also have an international context, as many end-users of federal biometric systems are foreign visitors. This realization is complicated as privacy expectations and experiences vary considerably throughout the world.

In December of 2005, the European Commission hosted a "Workshop on Ethical and Social Implications of Biometric Identification Technology: Toward an International Approach" in Belgium. Employees from OSTP and DHS' Privacy Office represented the U.S. government at this workshop. The primary purpose of the workshop was to initiate a dialogue between U.S. and Europe on biometric data protection and to compare laws, regulations, and social conventions.

The U.S. hosted a follow-on workshop in November 2006 called "International Conference on Biometrics and Ethics." This workshop brought together approximately 80 experts from several countries to engage in an open discussion of the application and ethics of biometrics. Participants included representatives from academia, private industry, non-profit organizations, and government, hailing from Asia, Europe, the Middle East, and North America. The workshop had four main panel discussions.

- Privacy & Ethics under Normal & Extraordinary Circumstances
- Ethics of Medical and Health Risks
- Ethics of International Data Sharing
- Government-Industry Collaboration

Communications

For the majority of the public, biometrics remains a technology they are more familiar with due to science fiction movies than practical experience. This lack of familiarity impacts perceptions of both the technology and its application by government agencies. Government communications activities have been aimed to not only enable the general public to better understand the technology and its applications, but also to educate the biometrics community on federal programs and plans so that they can better partner with the government to meet critical mission requirements.

> **"Promoting a scientifically educated and aware public is necessary if we are to make the appropriate decisions about the nation's R&D investments, guide the adoption and debate the societal implications of new science and technologies, and reap the maximum benefits from our investments. The quality of these efforts underpins the entire US scientific enterprise."**
>
> --Science for the 21st Century, July 2004

1. Communications Group

In 2006, the NSTC Subcommittee on Biometrics and Identity Management established an *ad hoc* interagency communications group to develop and coordinate an integrated outreach strategy for the government. The group's primary objective was to ensure an accurate awareness and understanding of biometric technologies and federal programs by the public, press, and Congress. Representative activities include:

- consistent use of key biometric terms throughout the government;
- coordination and messaging on outreach activities (such as conference presentations, press inquiries, etc.); and
- enhanced public websites and liaison activities.

Some specific activities are described in more depth below.

2. Biometric Consortium Conference

The Biometric Consortium's charter was established and formally chartered in 1995 by a committee working under authority of the Security Policy Board under Presidential Decision Directive/NSC-29. The Security Policy Board was subsequently abolished with issuance of National Security Presidential Directive 1, signed in February 2001, and the Biometric Consortium activities were eventually integrated with the communications activities of the NSTC Subcommittee on Biometrics and Identity Management. Today, the Biometric Consortium's primary function is to organize and host an annual conference, which serves as the federal government's major outreach effort each year. During this conference, government agencies openly discuss past activities and future plans and exchange ideas and lessons learned with the industrial and academic communities, as well as with foreign partners. Conference participation has skyrocketed since 2000 with three times as many attendees. The three-day conference now has three concurrent tracks, nearly 100 exhibitors, and approximately 2,000 participants from around the world, easily making it the world's premier biometrics-based conference.

3. Foundation Documents

In 2006, the NSTC Subcommittee on Biometrics and Identity Management released a series of foundation documents to enable the public to obtain an accurate understanding of biometric technologies and issues. Individual papers provided a top-level overview in an easy-to-understand manner while technology-specific papers provided a further level of specificity.

This set of documents also included a Biometrics Glossary, which represented government-wide consensus on terminology and definitions that agencies would use in their outreach material to make them more consistent across agencies and easier to understand.

The full set of foundation documents is available at http://www.biometrics.gov/ReferenceRoom/Introduction.aspx.

4. Websites

The NSTC Subcommittee on Biometrics and Identity Management hosts a triad of websites to help disseminate biometric information to the public.

- Biometrics.gov is the central source of information on biometrics-related activities of the federal government. Visitors to this site will find general information about biometrics and interagency collaboration activities, as well as introductions to federal biometric programs.
- The Biometrics Catalog is a user-updated repository of biometrics-related public information. The Biometrics Catalog is a searchable database of biometrics documents such as government reports, commercial products, evaluation reports, news articles, and a calendar of events.
- The Biometric Consortium Website provides information about its annual conference and hosts a bulletin board for public discussion of biometric technologies and issues.

These websites, working together, were developed to encourage greater collaboration and sharing of information on biometric activities among government departments and agencies; commercial entities; state, regional, and international organizations; and the general public.

5. International Meeting of Biometrics Experts

NIST hosted the International Meeting of Biometrics Experts on March 23-25, 2004, in Gaithersburg, Md., to discuss international sharing of testing protocols and approaches in biometrics. The implementation of efficient and effective biometric systems on an international scale can be greatly aided through international sharing of testing and evaluation methodologies and protocols, test results, and approaches that can be used by different nations to test their own systems. The purpose of this unique meeting was to support technical information sharing on biometrics between the various national biometrics testing laboratories as well as other appropriate organizations in the hope of leading to coordination of testing and evaluation procedures for the biometric components of travel documents. The impetus for this meeting came from a G-8 high-level working group on biometrics that focused on biometrics for travel documents. This group recognized the need for additional research to improve the accuracy of biometric systems and for international experts to meet to exchange ideas on testing methodologies.

The meeting was organized jointly by the Department of Commerce, DHS, DOS, DOJ, and the OSTP. The meeting was attended by senior technical experts, including experts in biometric science and biometric systems testing and evaluation, border and immigration systems including database and information technology systems, and law enforcement, particularly international law enforcement. Presenters and attendees represented G-8 and other nations including Australia, Canada, European Commission, France, Germany, Hungary, Italy, Ireland, Japan, Korea, Malaysia, Mexico, Netherlands, Norway, Russia, Singapore, South Africa, Switzerland, United Kingdom, and United States. The high level of expertise representing the governments of so many countries made this meeting a uniquely important contribution to biometrics for border security and law enforcement.

Government-wide Coordination, Strategies and Policies

1. Coordination

Executive Order 12881, signed by President Clinton on November 23, 1993, established the National Science and Technology Council (NSTC) and directed it to:

- coordinate the science and technology (S&T) policy-making process;
- ensure S&T policy decisions and programs are consistent with the President's stated goals;
- help integrate the President's S&T policy agenda across the federal government;
- ensure S&T are considered in development and implementation of federal policies and programs; and
- further international cooperation in S&T.

The NSTC established its Subcommittee on Biometrics and Identity Management in 2002[4], and it has been coordinating biometrics S&T policy issues ever since. Coordination approaches and priorities have changed over the years, which is normal as technologies are advanced and issues evolve from pure science to the application of science to meet specific operational needs.

Phase 1 2002-2003	Phase 2 2003-2006	Phase 3 2006-Present
Goals: <ul><li>Share lessons learned from operational systems</li><li>Grow USG biometrics expertise</li><li>Build relationships</li></ul>	**Goals:** <ul><li>Advance technology, privacy & communications</li><li>Grow USG biometrics expertise</li><li>Build relationships</li></ul>	**Goals:** <ul><li>USG-wide biometric system of systems</li><li>Community able to meet other government and private sector needs</li><li>Expansion to IdM</li></ul>
Deliverables <ul><li>List of topics for potential collaboration</li><li>Initiate joint RDT&E efforts</li></ul>	**Deliverables** <ul><li>Joint RDT&E successes</li><li>Foundational documents</li><li>Privacy paper & websites</li><li>*The National Biometrics Challenge*</li></ul>	**Deliverables** <ul><li>Interoperable Systems</li><li>USG-wide plans for standards, RDT&E, privacy & communications</li><li>Enhanced operational capabilities</li></ul>

---

[4] The activity received its first formal charter in the spring of 2003 as an Interagency Working Group on Biometrics. It was elevated to subcommittee status in 2005, and its area of responsibility was expanded to include Identity Management in January of 2007.

Initial subcommittee activities focused on expanding knowledge of biometrics within the government, sharing operational lessons learned to assist in rapid deployment of biometric technologies, building interagency relationships, and identifying unmet operational needs. One of the first activities was the Government Biometrics Workshop[5] held in March of 2003 and hosted by DoD, DOJ, and Treasury. The goal of the workshop was a first cut at identifying common operational needs in order to develop an interagency roadmap for future RDT&E.

Having identified an initial roadmap, the Subcommittee quickly transitioned to a focus on rapid advancement of the technology, developing standards at the national and international level, advancing and performing evaluations, and promoting privacy protection. Numerous projects from this phase are described throughout this paper. A summation of current RDT&E priorities can be found in The National Biometrics Challenge.

The National Biometrics Challenge

National Science and Technology Council
Subcommittee on Biometrics

August 2006

By 2006, the technology had sufficiently advanced and interagency relationships had grown to the point that in-depth collaboration on future biometric systems and overall governance could begin. The Subcommittee, while continuing its core S&T mission, also stepped out to begin interagency collaboration of this piece under an *ad hoc* interagency Interoperability Working Group. In June, 2006, the President approved the National Implementation Plan for the War on Terror (NIP) and in October 2006

---

[5] The workshop's report is available at
http://www.biometricscatalog.org/gbw/2003_US_Government_Biometrics_Workshop-Overview_and_Summary_Report.pdf

the National Security Council tasked the National Counterterrorism Center (NCTC) to ensure that the diverse, varied, departmental efforts to employ biometric technology to meet counterterrorism objectives are harmonized, de-conflicted and efficiently implemented and focused on using biometrics to identify known and suspected terrorists (KSTs).

Representatives of NCTC and OSTP met soon after the NIP was signed to determine how to best coordinate efforts. A joint approach was agreed to, with NSTC continuing to have primary S&T responsibilities and the NCTC sponsoring an interagency coordination group (ICG) focusing on the operational planning and implementation roles. When policy recommendations had been developed, they would be approved by the NSTC for S&T issues and by the National Security Council for operational issues[6]. Several individuals worked on both the Subcommittee and the ICG, and the Interoperability Working Group reported to both because its tasking had equal parts technology and operational issues. The Subcommittee and the ICG worked together, mutually driving each other to a successful conclusion. In March 2008, the National Security Council (NSC) Deputies Committee approved a strategic framework that summarized a number of recommendations developed by the Subcommittee and the ICG and directed agencies to begin planning their implementation.

In June 2008, President Bush signed NSPD-59/HSPD-24, "Biometrics for Identification and Screening to Enhance National Security." This directive builds upon the decisions in the strategic framework and initiates a process to expand operational coordination to national security concerns beyond KSTs.
Taken together, these activities represent the full cycle of government technology: identifying operational needs, which drives technology advancement, which serves as a foundation for policy/strategy formulation and implementation.

Looking even more broadly, biometrics is a subset of identity management, which is a topic that has emerged rapidly in recent years. Biometrics, badges and tokens of all kinds, passwords and personal identification numbers (PINs), etc., are just some of the current physical components of identity management. To all this must be added a wealth of law, regulation, policy, and awareness of and sensitivity to the attitudes and views of the organizations within which these systems are installed and operated. The Subcommittee chartered a subordinate NSTC Task Force on Identity Management in January 2008 to develop an initial roadmap for coordinated RDT&E activities, much like biometrics coordination efforts began six years prior.

2. Strategies

---

[6] National Security Presidential Directive 1 describes the working structure of the interagency National Security Council.

Prior to the issuance of NSPD-59/HSPD-24, a number of national strategies also discussed the importance of biometric technologies to support operational requirements. Some examples are listed below.

a. **The National Strategy for Maritime Security, September 2005, Executive Office of the President**

"In cooperation with the private sector, the United States will establish a system-wide common credential for use across all transportation modes by individuals requiring unescorted physical access to secure, restricted, and critical areas of the maritime domain. The identification card for access will use biometrics to link the person to the credential definitively."

"The rapid and accurate identification of individuals for access to secure, restricted, and critical areas is a paramount protection measure that must be implemented by the private sector in cooperation with the federal government. Persons seeking to enter the United States will undergo identity checks and biometric screening at the border and in the coastal approaches to verify their lawful admission."

b. **National Infrastructure Protection Plan, January 2006, DHS**

Access Control Systems (p. 34): Cyber systems allowing only authorized personnel and visitors physical access to defined areas of a facility. Access control systems provide monitoring and control of personnel passing throughout a facility by various means, including electronic card readers, biometrics, and radio frequency identification.

c. **National Strategy for Information Sharing, October 2007, Executive Office of the President**

Sharing Information with Foreign Partners (p. 25): "The counterterrorism mission requires sharing many types of terrorism-related information, for example, the exchange of biographic and biometric information related to known or suspected terrorists."

d. **National Strategy for Homeland Security, October 2007, Executive Office of the President, Homeland Security Council**

"In the face of resourceful terrorists... we must continue to expand the US-VISIT program's biometric enrollment from two fingerprints to ten fingerprints, as well as leverage science and technology to enable more advanced multimodal biometric recognition capabilities in the future that use fingerprint, face, or iris data."
"Create 'smart borders' (p. 22). We must prevent terrorists and the implements of terror from entering the United States. At the same time, our

economic security depends on the efficient flow of people, goods, and services. We will build a "smart border" that achieves both of these critical goals. It will feature strong, advanced risk-management systems, increased use of biometric identification information, and partnerships with the private sector to allow precleared goods and persons to cross borders without delay."

3. Policies

a. Budget Guidance Memorandum

FY2009 Administration Research and Development Budget Priorities Memorandum, August 2007, Office of Management and Budget (OMB) and Office of Science and Technology Policy (OSTP)
> "Rapid, reliable and accurate biometric-based recognition of individuals is necessary for successful homeland security, counterterrorism, border control, law enforcement, e-commerce and e-government, and identity theft prevention… As directed by the National Security Council's Deputies Committee, agencies are to place emphasis on the priorities outlined in The National Biometrics Challenge and the resulting agenda developed by the NSTC Subcommittee on Biometrics and Identity Management."

b. Presidential Directives

HSPD-6: Integration and Use of Screening Information, September 2003, Executive Office of the President.
> Provides for the establishment of the Terrorist Threat Integration Center, which became the NCTC.

HSPD-11: Comprehensive Terrorist-Related Screening Procedures, August 2004, Executive Office of the President.
> Implements a coordinated and comprehensive approach to terrorist-related screening that supports homeland security at home and abroad. This directive builds upon HSPD-6.

HSPD-12: Policy for a Common Identification Standard for Federal Employees and Contractors, August 2004, Executive Office of the President.
> Establishes a mandatory, government-wide standard for secure and reliable forms of identification issued by the federal government to its employees and contractors (including contractor employees).

NSPD-59/HSPD-24: Biometrics for Identification and Screening to Enhance National Security, June 2008, Executive Office of the President.
> Establishes a framework to ensure federal departments and agencies use compatible methods and procedures in the collection, storage, use, analysis, and sharing of biometric and associated biographic and

contextual information of individuals known or suspected to be a national security threat in a lawful and appropriate manner while respecting privacy and other legal rights under U.S. law.

Conclusion

The U.S. government studied and worked with biometrics long before 9/11, but the technology has experienced rapid growth and attention in the ensuing years by agencies and high-level coordination bodies. Working closely with the private sector and international partners over the past seven years, federal agencies have advanced the scientific basis of the technology and its mutual co-existence with fundamental privacy principles, rapidly implemented operational systems to meet immediate needs, and laid the foundation for maximizing the appropriate use of biometrics in future identity management applications.

The technological and operational growth of biometrics since 9/11, as well as its rise in stature within the current administration, has been unprecedented in the technology's history. Continued attention over the ensuing years will reap even greater operational capability while continuing to ensure privacy protection.

Appendix A – Expanded RDT&E Discussion

Before being thrust into wide-scale U.S. government operations, biometrics needed to develop a more stable scientific footing. The U.S. government therefore implemented a series of research, development, test and evaluation (RDT&E) and standards development activities, several of which are discussed below. These activities have significantly advanced the capabilities and our understanding of biometric technologies. These advancements have enabled the establishment and enhancement of many government systems in use today, such as those that screen for KSTs, while simultaneously maintaining personal privacy and civil liberties.

a. HumanID

DARPA's Human Identification at a Distance (HumanID) Program began in September 2000. The goal at that time was to develop automated, multimodal, multi-biometric surveillance systems for identifying humans at a distance for protection and for early warning against asymmetric threats. The state-of-the-art capability at that time on cooperative subjects, indoors, with controlled illumination was less than 10 feet. By the end of the program in 2003, some technology was capable of recognizing people at up to 150 feet. However, there was much to be done to improve performance.

The HumanID program provided the scientific foundation for human identification at a distance across the board. Various types of biometric technology were explored, to include face recognition, iris recognition, Doppler radar, infrared imagery, physiological methods such as pulse and heartbeat, and gait (recognizing someone by their walk). The HumanID Gait Challenge was the first time the potential of gait as a biometric had been thoroughly investigated. The HumanID program also developed the first prototype system for recognizing iris at a distance.

Overall, the HumanID program made significant gains in understanding the difficulties associated with biometric technology and provided the groundwork for future biometric technology programs.

b. Face

i. Face Recognition Vendor Tests

The Face Recognition Vendor Tests (FRVTs)[7] were a series of independently administered technology evaluations for face recognition systems. As face recognition technology began to be commercialized in the late 1990s, the government needed a way to measure the performance accuracy of these systems. The FRVT evaluation was created as a means

---

[7] http://www.frvt.org

of measuring the state-of-the-art performance capabilities and providing a methodology for evaluating this technology. Since that time, there have been three FRVT evaluations. Each successive evaluation increased in size, difficulty, and complexity. All of these evaluations were jointly sponsored by multiple federal government agencies.

After 9/11, the biometrics industry as a whole was thrust into the foreground of government and commercial operational requirements. Though face recognition technology had been in the commercial industry for a few years, 9/11 highlighted the importance of including biometrics in security applications. This launched a new interest in fielding face recognition systems to solve operational security requirements. There became an immediate need to expedite technology development and an urgency to assess the current state-of-the-art capabilities of commercial biometric technology. A new independent technology evaluation was required to determine if face recognition technology could answer this call. The first evaluation for face recognition after 9/11 was the Face Recognition Vendor Test (FRVT) 2002.

The FRVT 2002 was sponsored by six different U.S. government organizations, including DARPA, DOS, the National Institute of Justice (NIJ), NIST, FBI, and TSA.

The primary objective of the FRVT 2002 was to assess the capability of mature automatic face recognition systems to meet real-world applications. Achieving this objective required an evaluation that was much larger and broader in scale than the previous biometric evaluations. The increase in scale included the number of individuals in the evaluation as well as the detail and depth of analysis performed. It also required designing a new biometric evaluation protocol and establishing a new standard for evaluations.

At the end of the FRVT 2002, face recognition performance for verification had improved significantly. The verification error rate on full-frontal face images taken indoors with controlled illumination decreased from 79 percent (c. 1993) to 20 percent.

The FRVT 2002 showed it was possible to conduct large-scale biometric technology evaluations with greater than 100,000 biometric samples. Because the FRVT 2002 was so successful, large-scale evaluations are now routinely conducted in face, fingerprint, and iris recognition.

The onset of the war in Iraq brought an even greater need for deployable face recognition systems. It was the first time biometric technology was in such high demand for operational scenarios. To meet this demand, more expedient technology development and evaluation methodology was required.

The challenge problem and evaluation methodology for rapidly advancing the performance of biometric technology began with the Face Recognition Grand Challenge (FRGC) technology development program and the FRVT 2006 independent evaluation.

ii. Face Recognition Grand Challenge

The Face Recognition Grand Challenge (FRGC)[8] and FRVT 2006 established a new paradigm in computer vision for rapidly improving a technology's performance. The FRGC began in May 2004 and was the first technology development program of its kind for face recognition technology. The main objective of the FRGC was to improve face recognition verification performance by an order of magnitude over the FRVT 2002 results. The FRGC structure consisted of a set of challenge problems developed by the FRGC evaluation team, the test data, and the test software infrastructure. Each challenge problem consisted of a set of experiments designed to guide technology development to meet U.S. government operational requirements.

To emulate operational data, much more test data were needed. The FRGC collected one of the largest repositories for face recognition test data in the world. The test data were collected in strict adherence with Institutional Review Boards and all subjects signed consent forms prior to data collection. The test data consist of two-dimensional still images, both indoors and outdoors, and three-dimensional imagery. These test data were distributed to participating face recognition researchers and developers through a series of workshops. Participants were given the challenge problems, the infrastructure for conducting experiments, and the test data on which to develop their algorithms. Participants were asked to submit their results to the FRGC program manager for compilation and to present their results at subsequent workshops. Participants' self-reported results were used as the first-level performance gage to determine if the goal of an order of magnitude improvement was met. The results showed that it had indeed achieved its goal. However, independent verification of this achievement was required in order to make the claim. The FRVT 2006 did just that—confirmed the goal had been met.

The main objective of the FRVT 2006 was to assess the current state-of-the-art performance of face recognition technology and determine if it met

---

[8] http://face.nist.gov/frgc/

the FRGC goal of improving the verification error rate by an order of magnitude over the FRVT 2002 results.

The FRVT 2006 documented significant progress since January 2005 in face recognition when faces were matched across different lighting conditions. In the FRVT 2006, an evaluation on sequestered data, five submissions performed better than the best results in the January 2005 FRGC results. The observed increase occurred despite the FRGC being an open challenge problem with the identities of faces known to the FRGC participants and the FRVT 2006 being an evaluation on sequestered data.

The FRVT 2006 and FRGC programs were also the first to include human performance versus machine performance testing. FRVT 2006 integrated human face recognition performance into an evaluation for the first time. This inclusion allowed a direct comparison between humans and state-of-the-art computer algorithms. The study focused on recognition across changes in lighting. The experiment matched faces taken under controlled illumination against faces taken under uncontrolled illumination. The results show that, at low false alarm rates for humans, seven automatic face recognition algorithms were comparable to or better than humans at recognizing faces taken under different lighting conditions. Furthermore, three of the seven algorithms were comparable to or better than humans for the full range of false alarm rates measured.

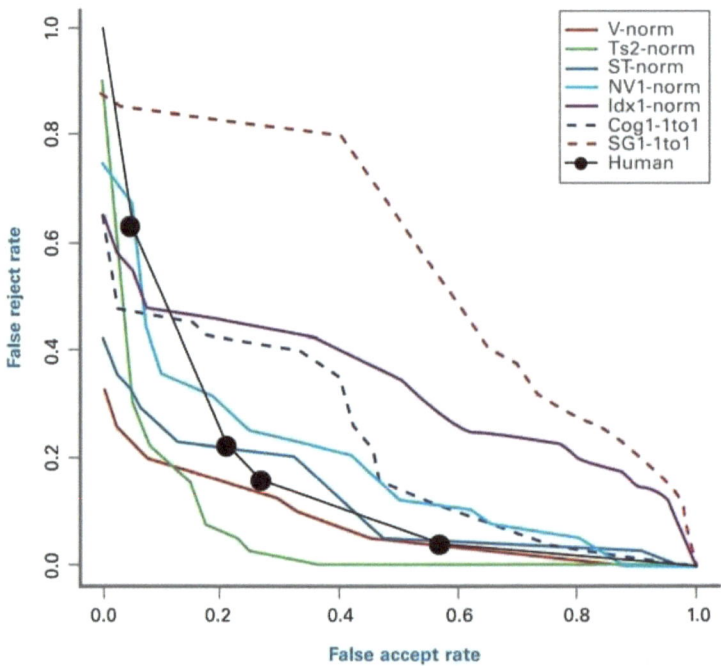

Figure 1. ROC of human and computer performance on matching faces across illumination changes.

At the conclusion of the FRVT 2006, results confirmed that the goal had been met. Face recognition verification performance for frontal face stills in controlled illumination had improved by an order of magnitude over FRVT 2002 results, from a 20 percent error rate to a less than 1 percent error rate [*Figure 2*]. In fact, face recognition performance had improved by two orders of magnitude since the beginning of the FERET program in 1993. This rapid improvement can partly be attributed to the technology development and evaluation methodology established by NIST and its sponsors.

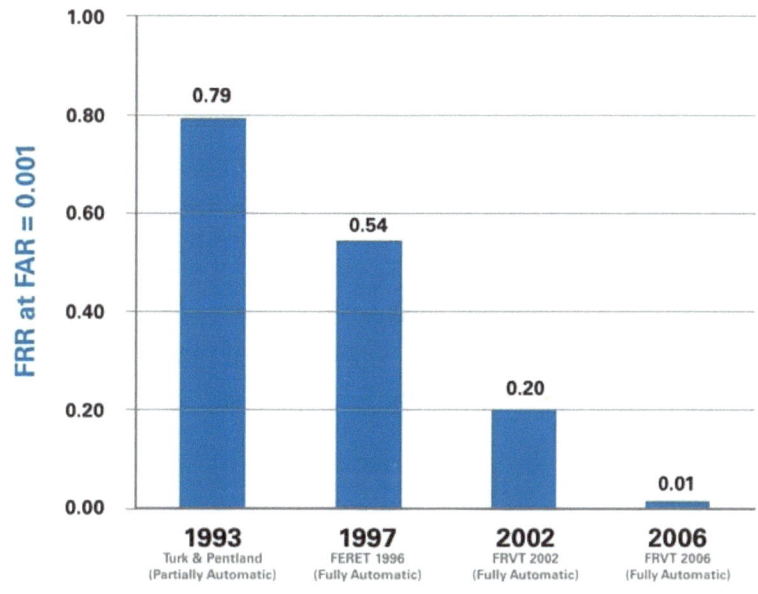

Figure 2. The reduction in error rate for state-of-the-art face recognition algorithms as documented through the FERET, the FRVT 2002, and the FRVT 2006 evaluations.

The FRVT 2006 and FRGC were jointly sponsored by six U.S. government agencies and organizations, to include the DHS' Science and Technology Department and TSA, the Director of National Intelligence's Information Technology Innovation Center, the FBI, the NIJ, and the Technical Support Working Group. The test data collected by the FRGC program are still being requested and used worldwide to further develop face recognition technology algorithms and improve performance.

While the FRVT 2006 and FRGC worked well to advance face recognition technology, they also highlighted areas requiring more research to meet new operational requirements. These areas include imagery taken in uncontrolled conditions, such as hallways or outdoors, and video. These areas of interest form the foundation for the follow-on program, the Multiple Biometric Grand Challenge, which is described below.

c. Finger

iii. FpVTE

The Fingerprint Vendor Technology Evaluation (FpVTE) 2003[9] was conducted to evaluate the accuracy of fingerprint matching, identification, and verification systems. FpVTE was conducted by NIST on behalf of DOJ's Justice Management Division (JMD). FpVTE serves as part of the NIST statutory mandate under section 403(c) of the USA PATRIOT Act to certify biometric technologies that may be used in the US-VISIT Program.

The FpVTE evaluations were conducted to:
- measure the accuracy of fingerprint matching, identification, and verification systems using operational fingerprint data;
- identify the most accurate fingerprint matching systems;
- determine the effect of a wide variety of variables on matcher accuracy; and
- develop well vetted sets of operational data from a variety of sources for use in future research.

Planning for FpVTE started in May 2003, and analysis continued through April 2004. Eighteen different companies participated, with 34 systems tested. Participants were required to assemble, configure, and run their own hardware and software at NIST's Gaithersburg, Md., facility. The trials began in October 2003, with each participant running over a two- or three-week period depending on which of the three specific tests they participated in and using a predetermined and staggered schedule. Testing of all 18 different companies was completed in November 2003.

At the time, FpVTE 2003 was the most comprehensive independent evaluation of fingerprint matching systems ever executed, particularly in terms of the number and variety of systems and fingerprints. More than 48,000 sets of operational-quality fingerprints from more than 25,000 individuals were used in FpVTE.

Conclusions from FpVTE include the following.

- The top-performing systems performed consistently well over a variety of image types and data sources.

---

[9] Charles Wilson, R. Austin Hicklin, Harold Korves, Bradford Ulery, Mellisa Zoepfl, Mike Bone, Patrick Grother, Ross Micheals, Steve Otto, & Craig Watson, "Fingerprint Vendor Technology Evaluation 2003: Summary of Results and Analysis Report," NISTIR 7123, June 2004, http://fpvte.nist.gov/.

- These systems produced matching accuracy results that were substantially different than the rest of the systems.
- The variables that had the largest effect on system accuracy were the number of fingers used and fingerprint quality:
    - Additional fingers greatly improve accuracy.
    - Poor quality fingerprints greatly reduce accuracy.
- Capture devices alone do not determine fingerprint quality.
- Accuracy can vary dramatically based on the type of data.

iv. Proprietary Fingerprint Template (PFT) testing

Since June of 2003, NIST has been conducting tests of fingerprint-based biometric matching systems using vendor-supplied software development kits (SDKs). This testing program has been named Proprietary Fingerprint Template (PFT) testing[10] because vendors submit matching algorithms that are permitted to use any potentially proprietary fingerprint features they determine useful. (This is in contrast to the interoperable fingerprint template testing (MINEX) described later in this section.) Fingerprint matching algorithms from vendors are being evaluated to insure that the accuracy of the matchers used in various existing and planned government systems (including FBI/IAFIS and DHS/US-VISIT) are comparable to the most accurate available COTS products. PFT measures the state-of-the-art in one-to-one matching for verification over a wide range of fingerprint image qualities.

In PFT testing, an application written by NIST controls calls to two vendor-supplied SDK functions. The first function (extraction) is used to create the fingerprint matcher templates, and the second function (matcher) compares two fingers at a time and returns a match score. All testing is performed by NIST personnel and run on NIST computer hardware.

Performance was originally reported on matching single index fingers. NISTIR 7221 "Studies of One-to-One Fingerprint Matching with Vendor SDK Matchers"[11] reports results on earlier single-finger matching. The evaluation of fingerprint SDK matchers was extended to evaluate the matching accuracy that can be achieved by combining scores for the right and left index fingers to support work at NIST on Personal Identity Verification (PIV)[12] for HSPD12[13]. NISTIR 7249

---

[10] See PFT Testing Homepage at http://fingerprint.nist.gov/pft/

[11] Craig Watson, Charles Wilson, Karen Marshall, Mike Indovina, & Rob Snelick, "Studies of One-to-One Fingerprint Matching with Vendor SKD Matchers," NISTIR 7221, April 2005.

[12] See NIST PIV Project at http://csrc.nist.gov/groups/SNS/piv/

[13] See Homeland Security Presidential Directive (HSPD) 12 at http://csrc.nist.gov/drivers/documents/Presidential-Directive-Hspd-12.html

"Two Finger Matching with Vendor SDK Matcher"[14] reports results on two-finger matching.

The PFT testing program continues to this day. Vendors are permitted and encouraged to submit their latest and greatest algorithms for evaluation. In this way, the progression of the state of the art of fingerprint matchers can be tracked. To date, more than 15 organizations have participated in PFT and, in all, NIST has evaluated more than 30 different fingerprint matching algorithms.

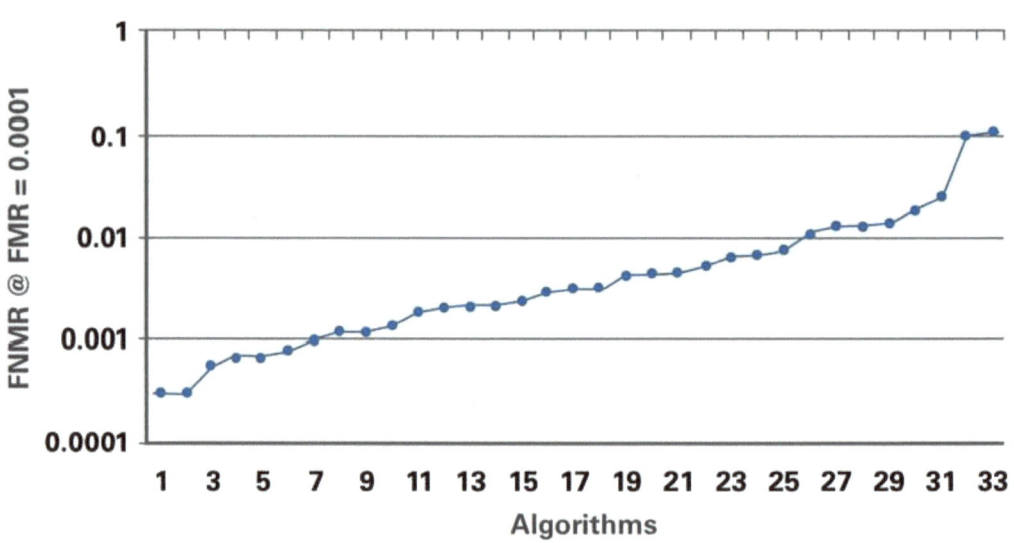

Figure 3. PFT error rates for two-finger matching on an operational dataset (POEBVA).

[14] Craig Watson, Charles Wilson, Michael Indovina, & Brian Cochran, "Two Finger Matching With Vendor SKD Matchers," NISTIR 7249, July 2005.

To summarize current PFT status,

PFT Two-Finger Matching

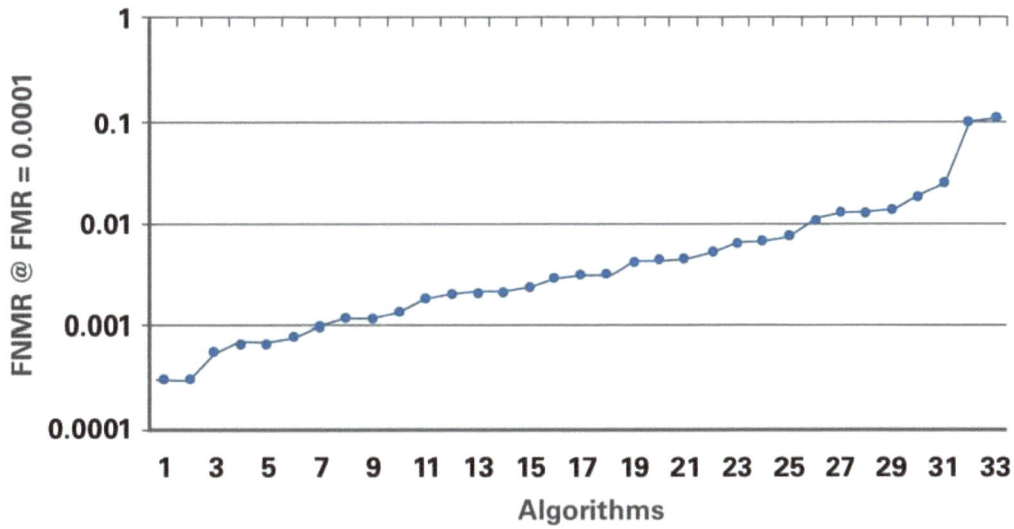

Figure 3 plots the False Non-Match Rate (FNMR) measured at a constant False Match Rate (FMR) of 0.0001 (that's one false match in 10,000 matches) for two-finger matching on an operational dataset (POEBVA).

- The top seven matchers with lowest error rates are contributed by the three top-performing vendors in FpVTE;
  (FNMR in this group ranges from 0.0003 to 0.0010 at a FMR of 0.0001).
- There are two outliers of poor-performing systems;
  (FNMR of these two is greater than 0.0900 at an FMR of 0.0001).
- Algorithms in between these two groups steadily decline in performance;
  (FNMR ranges from 0.0010 to 0.0245).

The wide range of error rates plotted in this figure demonstrates the need for public tests such as PFT.

v. Slap fingerprint segmentation evaluation

The Slap Fingerprint Segmentation Evaluations at NIST assess the accuracy of algorithms used to segment slap fingerprint images into individual fingerprint images.

- **Slap fingerprints** are taken by simultaneously pressing the four fingers of one hand onto a scanner or fingerprint card. Slaps are also known as four-finger simultaneous plain impressions.
- **Slap segmentation** is the process by which a slap image (a four-finger simultaneous plain impression) is divided into four images of the individual fingers.

Slap fingerprints have the advantage of enabling much quicker capture of all 10 fingers than individually rolling each finger; however, slaps have the disadvantage of only being plain impressions, whereas rolled fingerprints collect more friction skin data (nail to nail).

SlapSeg 04[15] was conducted by NIST in 2004 on behalf of the DOJ/JMD, IDENT/IAFIS Integration Project. Additional partners included the US-VISIT Program of DHS and the FBI. At that time, the use of slap fingerprints for background checks was being considered in a variety of U.S. government fingerprint systems (including US-VISIT and IAFIS). Automated segmentation of slap fingerprints was known to have an associated error rate, but no rigorous evaluation of current slap segmentation algorithms had ever been conducted before SlapSeg 04. Knowing whether existing segmentation software was of sufficient accuracy for operational use was of practical interest and value to policymakers.

SlapSeg 04 was conducted to determine the accuracy of existing slap segmentation algorithms on a variety of operational-quality slap fingerprints. Segmentation algorithms were evaluated based on their abilities to:

- produce highly matchable images;
- identify finger positions; and
- detect segmentation failures.

They were evaluated using a variety of data from different sources. The fingerprints were acquired from inked paper cards (subsequently scanned) or by livescan devices. The images had a wide range of operational and non-operational quality.

Conclusions from SlapSeg 04 included the following.

- The most accurate segmenters produced at least three highly matchable fingers and correctly identified finger positions in from 93% to over 99% of the slap images, depending on the data source.
- The data source had a much greater effect on success rate than whether the images were collected using livescan devices or paper.
- Most segmenters achieved comparable accuracies on the better quality data, but there were significant differences among segmenters when processing poor quality data.

---

[15] Bradford Ulery, Austin Hicklin, Craig Watson, Michael Indovina, & Kayee Kwong, "Slap Fingerprint Segmentation Evaluation 2004 (SlapSeg 04) Analysis Report," NISTIR 7209, March 2005, http://fingerprint nist.gov/slapseg04/.

- Some segmenters are capable of identifying many, but not all, problem slaps; failure rates could be cut substantially by allowing some of the slaps to be recaptured or rejected.

Given the results of SlapSeg 04, federal agencies, including the FBI, DOS, and DHS, are now using slap fingerprints to support applications such as civilian background checks, U.S. visa issuance, and the US-VISIT Program. NIST has recently announced a new series of slap segmentation evaluations called SlapSeg II[16]. Unlike SlapSeg 04 (which relied on matchability as the criterion for successful segmentation), NIST intends for SlapSeg II to use a measure of success based the location and boundaries of the segmented fingerprints. This will reduce the time and labor needed to administer these tests, facilitating larger-scaled studies and enabling participants to submit multiple algorithms over time. This will serve to improve segmentation technology by fostering competition and innovation.

Through SlapSeg II, technology users will benefit from knowing how much the current state of the art in slap segmentation technology has advanced. Vendors will also benefit as they will gain the knowledge of how their segmentation implementation performs on a large collection of operational-quality data.

vi. Fast fingerprint slap capture

In the fall of 2005, DHS, FBI, DOS, NIJ, DoD/BFC, and NIST jointly defined an urgent, near-term demand for faster, smaller, more mobile, 10-fingerprint slap capture devices to meet critical national security needs. These departments organized a unified 10-Print Capture User Group to develop common requirements and co-sponsor a "Challenge to Industry" as a first step toward meeting these common needs.

The User Group identified the need for a 10-print capture scanner device (the "Scanner") along with client- and server-based utility software (the "Software"), including slap quality, slap segmentation, sequence verification, fingerprint image quality, compression, and other utilities. The Scanner and Software had to be interoperable; that is, any approved Scanner must work with any approved Software and vice versa.

The essential operational requirements of the Scanner and Software (some of which were not met by the industry at that time) included:

---

[16] SlapSegII Homepage @ http://fingerprint.nist.gov/slapsegII/

- meet the current space requirements (6"x6"x6") constraints of the deployment facilities;
- be mobile so as to support multiple operational scenarios;
- perform all the 10-print capture processing steps including individual finger segmentation and image quality checks within five seconds per slap from the time the subject places his/her fingers on the Scanner to the moment the capture Software has segmented, analyzed image quality, and delivered status to the operator;
- be powered without an additional 120v power plug in order to meet power capacity and power cabling constraints of the current facilities;
- comply with the current biometric industry standards; and
- meet or exceed fingerprint quality requirements contained in the latest version (7.1) of the FBI's Electronic Fingerprint Transmission Specification, Appendix F.

The size of the Scanner, power requirements, and speed of both the Software and Scanner were of critical importance due to their impact on facilities and operations.

In October 2005, NIST hosted the 10-Print Capture Scanner & Software Requirements Workshop.[17] At this workshop, the User Group announced its plan to conduct market research and presented industry with the challenge of developing a fingerprint scanner that met these requirements. A Request for Information (RFI) was issued, and a number of interested vendors submitted white papers in response telling the User Group how they planned to meet these requirements by bringing technology to market within one year. A selected set of vendors who responded to the RFI were then invited to meet with the User Group one-on-one to gain further insight into the state of the market and how firms intended on meeting the requirements within the necessary time frame.

At the end of its market research, the User Group concluded that, while no vendor at that time had a complete solution available, industry would be able to meet the general objectives and meet the User Group's core requirements in the expected time frame of 12 months.

Just over one year later, in November 2006, DHS hosted a 10-Print Capture User Group Industry Day. At this event, it was publicly announced that industry had stepped up to the challenge of the User Group's requirements, and industry was able to demonstrate functioning 10-print capture scanners and software that had not existed one year earlier.

---

[17] 10-Print Capture Scanner & Software Requirements Workshop, October 2005, documents found at http://www.itl.nist.gov/iad/894.03/pact/pact.html.

Today, 10-print slap capture fingerprint devices are being used within DOS BioVISA, DHS US-VISIT, and FBI IAFIS civilian background checks. The 10-Print Capture User Group's tremendous success serves as a model of how federal agencies can come together, form common procurement requirements, promote a unified market to industry, and have industry step up and deliver with great efficiency for the U.S. taxpayer.

vii. Fast rolled-equivalent fingerprint capture

The Fast Capture Rolled-Equivalent Finger/Palm Print Initiative began in January 2004 to improve and advance the current state of technology for the capturing of 10 rolled-equivalent fingerprints or fingerprints and palm prints. The resulting technology will provide the ability to capture 10 rolled-equivalent fingerprints in 15 seconds or less and both palms in 1 minute or less. Significantly greater convenience, speed, reliability, affordability, and accuracy for finger and palm capture will improve our nation's ability to meet the screening requirements for criminal, terrorist, border, transportation, and employment checks.

This initiative was a joint effort of the NIJ, FBI, Drug Enforcement Agency, and the JMD; and the DoD, DHS, and DOS. The agencies worked cooperatively to define the requirements, author a solicitation, review the applications, and fund four resulting projects. The NIJ took the lead on awarding and managing the projects.

Each project approached capturing the prints in a different way: a visual 3-dimensional model of the hands constructed from multiple camera images stitched together; analysis of structured-light interference patterns on the fingerprints; a 2-dimensional flexible polymer plastic foil sensor array; and a motion-controlled scanning three-camera assembly. Initial demonstrations of the technologies occurred in May 2007 showing that capturing 10 rolled-equivalent fingerprints in 15 seconds was feasible, and second phase efforts were funded in FY08 to continue to engineer the solutions into prototype devices. Concurrently, funding was provided to NIST to provide independent technology assessment of the prototypes and to develop evaluation criteria by which to certify their performance against national standards and data interchange formats.

Early concepts are already finding follow-on DHS support to evaluate contactless capture of fingerprints to support faster screening and reduce concerns of exposure to germs or contaminants on touch devices.

viii. Latent fingerprint testing

NIST is conducting a series of tests called Evaluation of Latent Fingerprint Technologies (ELFT)[18] for evaluating the state of the art in automated latent fingerprint matching. ELFT is being conducted for the DHS Science & Technology (S&T) Directorate and the FBI. The scope and structure of these tests are based partly on lessons learned from the April 2006 NIST Latent Fingerprint Testing Workshop[19], supplemented by technical interchanges with workshop participants and vendors. The intent of the testing is to quantify the core algorithmic capability of contemporary matchers.

While the immediate goal of ELFT is to assess automated technology, long-term goals go far beyond simply quantifying performance. It is fully expected that understanding the performance envelope and limitations of contemporary matchers will lead to improvements in technology. These in turn will lead to enhanced performance for searches of 10-prints and plain impressions against unsolved latent databases and watch lists. Equally important, technology improvements will provide law enforcement with the capability to search their unsolved latent fingerprints against 10-print files with greatly reduced effort.

ELFT is structured as a multi-year project. The first part of this project consists of two phases run in a "lights-out" environment. Phase I was completed in 2007 and represents a proof-of-concept test the main purpose of which was to demonstrate integrity of the software, including the evaluation test-bed itself. During Phase I the participants' software demonstrated:

- automated feature extraction from latent images;
- ability to match these features against enrolled 10-print backgrounds; and
- generation of candidate lists.

Each participant in Phase I received a personalized performance report on their algorithms. Only aggregate results were released to the public. In summary, 10 participants contributed a total of 16 SDKs for evaluation. Each SDK was subjected to 100 latent searches against a background of 1,000 10-prints. The performance across algorithms varied greatly, with best performers achieving a lights-out latent hit rate at rank-1 of better than 80%, while the poorest performers achieved a hit rate of less than

---

[18] See NIST ELFT Project at http://fingerprint.nist.gov/latent/.
[19] Vladimir Dvornychenko & Michael Garris, "Summary of NIST Latent Fingerprint Testing Workshop," NISTIR 7377, November 2006.

30%. This wide range of performance demonstrates the need for continued public testing and technology improvements.

Phase II is currently under way and employs a larger database to quantify the achievable performance ("hit rate") for automated searches. In Phase II, each SDK is being subjected to 1,000 latent searches against a background of 10,000 10-prints. The performance of each SDK will be publicly disclosed in Phase II.

ix. Fingerprint minutiae interoperability testing

The Minutiae Interoperability Exchange (MINEX) family of tests is being conducted by NIST to support the use of fingerprint minutiae templates as the *de facto* leading biometric data element for large-scale identity management applications. This is centered primarily on storage of minutiae records on identity credentials but extends to the transmission of compact fingerprint data over operational and bandwidth-limited networks.

The MINEX program was established to determine the feasibility of using minutiae data (rather than image data) as the interchange medium for fingerprint information between different fingerprint matching systems. The key focus is on standardized minutiae data to achieve effective and efficient interoperability. The MINEX program is currently made up of the three tests described below: MINEX 04[20], Ongoing MINEX[21], and MINEX II[22]. Further activities in this area, MINEX III, IV, etc., are now being planned to enhance conformity, accuracy, and interoperability of minutiae-based systems. Federal partners who benefit from this program include DHS S&T, DHS US-VISIT, DOJ/JMD IDENT/IAFIS, and DOJ/FBI. Globally, the MINEX program has been influential on various identity management programs, including those for border management.

MINEX 04

The approval of the INCITS 378 fingerprint template standard creates the possibility of a fully interoperable multivendor marketplace for applications involving fast, economic, and accurate interchange of compact biometric templates. MINEX 04 was conducted from August 2004 through March 2006 to address the outstanding questions surrounding the new standard.

---

[20] Patrick Grother, Michael McCabe, Craig Watson, Mike Indovina, Wayne Salamon, Patricia Flanagan, Elham Tabassi, Elaine Newton, & Charles Wilson, "MINEX – Performance and Interoperability of the INCITS 378 Fingerprint Template," NISTIR 7296, March 2006, http://fingerprint nist.gov/minex04/.

[21] See NIST Ongoing MINEX Project at http://fingerprint nist.gov/minex/.

[22] Patrick Grother, Wayne Salamon, Craig Watson, Michael Indovina, & Patricia Flanagan, "MINEX II – Performance of Fingerprint Match-on-Card Algorithms Phase II Report," NISTIR 7477, February 2008, http://fingerprint nist.gov/minexII/.

- Does the template give accuracy comparable with proprietary (image-based) implementations?
- Can template data be generated and matched by different vendors without an increase in error rates?

The MINEX 04 evaluation was designed to answer these questions and compared proprietary templates against two variants of the INCITS 378 format – MIN:A, which codes minutiae (x, y, θ, type, quality) and MIN:B, which supplements it with ridge count, core, and delta information.

Fourteen vendors participated. All of them implemented the MIN:A template; six elected to implement the MIN:B enhancement; and each was baselined against its corresponding proprietary template technology. By using very large-scale trials and four archived operational datasets, conclusions from MINEX 04 included the following.

- Proprietary templates are superior to MIN:A templates in terms of lower error rates.
- The reduced accuracy obtained using standard templates compared to proprietary templates can be adequately compensated for by using two fingers for all authentication attempts.
- Some template generators produce standard templates that are matched more accurately than others; some matchers compare templates more accurately than others. The leading vendors in generation are not always the leaders in matching, and vice-versa.
- Certification of an interoperable group of products requires some prior specification of the required accuracy. Large numbers of products will interoperate when the accuracy requirement is low. Fewer vendors are interoperable in high-performance interoperability scenarios.

In terms of impact, the results of MINEX 04 were used in decisions for projects such as PIV. In response to MINEX 04, NIST released FIPS-201[23] in February 2005, which defines the structure of an identity credential. It specified the inclusion of data from two fingerprints as a third authentication factor. The format for this information was finalized in February 2006 when NIST Special Publication 800-76-1[24] specified essentially the MINEX MIN:A template as a profile of the INCITS 378 standard. The result of this is the presence of INCITS 378 templates in PIV cards to be carried by all employees and contractors of federal agencies. Other programs such as TSA's

---

[23] FIPS PUB 201: Personal Identity Verification (PIV) of Federal Employees and Contractors, March 2006, http://csrc.nist.gov/publications/fips/fips201-1/FIPS-201-1-chng1.pdf.
[24] Charles Wilson, Patrick Grother, & Ramaswamy Chandramouli, NIST Special Publication 800-76-1: Biometric Data Specification for Personal Identity Verification," January 2007, http://csrc.nist.gov/publications/nistpubs/800-76-1/SP800-76-1_012407.pdf.

Transportation Worker Identification Credential (TWIC) and Registered Traveler may adopt this specification, and together these biometric-enabled credentials will soon number in the millions.

Ongoing MINEX

Ongoing MINEX follows the approach of MINEX 04 and is a continuing evaluation of INCITS 378 fingerprint template interoperability. The test program has two mandates:

- to provide measurements of performance and interoperability of core template encoding and matching capabilities to users, vendors, and interested parties; and
- to establish compliance for template encoders and matchers for the U.S. government's PIV program.

The Ongoing MINEX program evaluates template encoding and matching software submitted to NIST in the form of an SDK library. This involves, at a minimum, the submission of an SDK that provides functionality to create MINEX-compliant templates based on individual fingerprint images. Participants are encouraged to also provide a template-matching function.

Participants in the Ongoing MINEX test may optionally submit their products to establish PIV compliance in accordance with section 7.4.1 of NIST Special Publication 800-76-1. Upon completion of testing, if the submitted SDK meets the performance criteria defined by NIST (the interoperable group of algorithms must maintain FMR = 0.01 and FNMR $\leq$ 0.01), it shall be considered MINEX-compliant and listed on the NIST website[25]. The effect is to establish a baseline for the core algorithmic capability of the providers' implementations. To date, 21 feature extractors (template generators) and 19 matching algorithms have been tested to be compliant. The program is being used to support credentialing efforts worldwide.

MINEX II

MINEX II was conducted between July 2007 and February 2008 to evaluate the accuracy, speed, and interoperability of Match-on-Card verification algorithms. Match-on-Card is an example of a privacy-enhancing technology in that a cardholder's biometric data never leave the host card. Instead, the template-matching calculation is executed on ISO/IEC 7816 smartcards. They compare conformant reference and verification instances of the ISO/IEC 19794-2 Compact Card fingerprint minutiae standard and render a verification decision. MINEX II was designed to answer the longstanding question of whether such complex matching algorithms running on smartcards can achieve accuracy approaching that of server-based algorithms. The MINEX II

---

[25] See NIST Ongoing MINEX PIV Compliance Website at http://fingerprint.nist.gov/minex/qpl.html.

test therefore represents an assessment of the core viability of the *de facto* leading compact biometric data element on personal identity credentials based on the industry standard smart card. The results are relevant to users seeking to use minutiae templates as an additional factor for authentication.

MINEX II did not evaluate interface standards, secure transmission protocols, or card or algorithm vulnerabilities. In addition, it did not mimic a particular verification scenario, and it did not compare fingerprint sensors or system-on-card implementations. MINEX II was conducted concurrently with a separate NIST activity, SBMOC[26], which was designed to assess feasibility of conducting cryptographically secure, contactless biometric authentication in less than 2.5 seconds. Participation in SBMOC was not required for participation in MINEX II.

The significant results from MINEX II include the following.

- The most accurate Match-on-Card implementation achieves the minimum error rate specifications of the U.S. government's PIV program.
- For the one provider who has submitted both Match-on-Card and Match-off-Card minutiae-matching algorithms to NIST, the accuracy of the former approaches that of the latter.
- The most accurate Match-on-Card implementation executes 50% of genuine ISO/IEC 7816 VERIFY commands in 0.54 seconds (i.e., median), and 99% within 0.86 seconds. For the fastest implementation, these values are 0.18 and 0.48 seconds, respectively.
- MINEX II attained unprecedented transparency in its execution; the evaluation plan was published during its development with industry, and version-controlled open-source software was released for both conformance and conversion of INCITS 378 and ISO/IEC 19794-2 Compact Card templates and for invocation of ISO/IEC 7816 Match-on-Card operations.

In June 2008, the national bodies of ISO/IEC JTC 1 Subcommittee 37, Biometrics, voted to initiate standardization of a match-on-card test protocol as ISO/IEC 19795 Biometric Performance Testing and Reporting – Part 7: Testing of ISO/IEC 7816-based verification algorithms. The U.S. national body has contributed the core of the MINEX II test plan, NISTIR 7485[27], toward a base working draft of the standard.

---

[26] David Cooper, Hung Dang, Philip Lee, William MacGregor, & Ketan Mehta, "Secure Biometric Match-on-Card Feasibility Report," NISTIR 7452, http://csrc.nist.gov/publications/nistir/ir7452/NISTIR-7452.pdf.

[27] Patrick Grother & Wayne Salamon, "MINEX II – Performance of Fingerprint Match-on-Card Algorithms Evaluation Plan," NISTIR 7485, August 2007, http://fingerprint nist.gov/minexII/nistir_7485.pdf.

d. Iris

The U.S. government funded iris recognition research for several years prior to 9/11. While iris recognition technology garnered acceptance as a highly accurate biometric, it required a high degree of cooperation from the user. To improve the utility, performance, and ease-of-use of this technology, the U.S. government increased its investment after 9/11. Notable advancements that can be attributed to this investment and foresight include but are not limited to: increased standoff distances and useable volume; increased system performance while reducing size and cost; and the demonstration of prototypes capable of acquiring and matching the iris of subjects while moving through a portal. In addition, the U.S. government has sponsored the development of multiple match algorithms, including government-owned.

Several elements of the U.S. government collaborated to sponsor the NIST Iris Challenge Evaluation, described in further detail below.

Other areas of influence include the sponsorship of academic programs to create U.S. experts and spawn new technologies that encourage commercial competition and foster the rapid introduction of technological advancements. In addition, the U.S. government has sponsored the development of multiple analysts' tools that augment automated iris match algorithms to address the needs of a broad array of government, industry, and academic partners. This influence has significantly advanced the state of the art and emphasizes interoperable iris biometric technology.

ICE

The technology development and evaluation methodology for face recognition worked very well on the FRVT 2006 and FRGC programs, improving face recognition performance by an order of magnitude over FRVT 2002 results. Some of the sponsors were also interested in doing the same type of technology development and evaluation for iris recognition. The Iris Challenge Evaluation (ICE) Program began in 2005. The ICE 2005 was the technology development phase of the ICE Program while the ICE 2006 was the evaluation phase.

As with the FRGC, the ICE 2005 consisted of a challenge problem, the test data, and the test software and infrastructure. To use the existing infrastructure, the ICE 2005 and ICE 2006 ran concurrently with the FRVT 2006 and FRGC. This was the first time, however, that iris recognition had been independently tested by one sensor on multiple algorithms. Results from the ICE 2006 were published with the FRVT 2006 results, since they ran concurrently on the same test infrastructure, and provided a baseline for future evaluations.

e. Biometric Quality

Performance of biometric systems depends on the quality of the acquired input samples. Accuracy of current biometric systems is high when high-quality samples are being compared. Performance, however, degrades substantially as quality drops. Although only a small fraction of input data are of poor quality, the bulk of recognition errors can be attributed to poor-quality samples. Poor-quality samples decrease the likelihood of a correct verification and/or identification, while extremely poor quality samples might be impossible to verify and/or identify. If quality can be improved, whether by sensor design, user interface design, or standards compliance, better performance can be realized. For those aspects of quality that cannot be designed in, an ability to analyze the quality of a live sample is needed. This is useful primarily in initiating the reacquisition from a user but also for the real-time selection of the best sample and the selective invocation of different processing methods. That is why quality measurement algorithms are increasingly deployed in operational biometric systems. With the increase in deployment of quality algorithms, the need to standardize an interoperable way to store and exchange biometric quality scores increases.

Biometric quality analysis is a technical challenge because it is most helpful when the measures reflect the performance sensitivities of one or more target biometric matchers. NIST addressed this problem in August 2004 when it issued the NIST Fingerprint Image Quality (NFIQ) algorithm. NFIQ is a fingerprint quality measurement tool. It is implemented as open-source software and is used today in U.S. government and commercial deployments. Its key innovation is to produce a quality value from a fingerprint image that is directly predictive of expected matching performance and has been designed to be matcher-independent. There is now international consensus in industry, academia, and government that a statement of a biometric sample's quality should be related to its recognition performance. Since its release, NFIQ has won national and international acceptance and has become the *de facto* standard. NFIQ is included in the Electronic Biometric Transmission Specification (EBTS), which is a required standard for doing business with the FBI's IAFIS.

Since 2004, NIST has been considering how quality measures should be evaluated, developing quality measures for other biometrics, and considering the wider use of quality measures in biometric systems, including quality summarization and quality calibration. In addition, NIST is active in SC 37 and M1 standardization activities on biometric quality and sample conformance.

The NIST Biometric Quality Program focuses on standards, tools, guidance, and workshops.

Standards

In January 2006, the SC 37 Biometrics Subcommittee of JTC 1 initiated work on ISO/IEC 29794, a multipart standard establishing quality requirements for fingerprint (Part 4), face (Part 5), generic aspects (Part 1), and possibly other biometrics later. Both DHS and FBI expressed a need for achieving interoperability of quality scores with other government agencies. The NIST Biometric Quality Program has contributed to the generic ISO quality draft

(ISO/IEC 29794-1) to require that quality values must be indicative of recognition performance; and NIST has made technical contributions on representation, storage, and exchange of quality scores. The goal is an improved standard that reflects the operational needs of the U.S. government, particularly DHS US-VISIT, TSA Registered Traveler, PIV, and the recent rollout of the international e-Passport.

Tools

As mentioned above, NIST has developed the free, open, and vendor-independent fingerprint image quality algorithm, NFIQ. NFIQ formalizes the concept of biometric sample quality as a scalar quantity that is related monotonically to the performance of biometric matchers under the constraint that at least two samples with their own qualities are being compared.

Guidance

NIST has published a technical contribution and guidance toward quality summarization. Quality summarization addresses the important issue of enterprise quality-assurance surveying by providing tools on how to combine quality scores of individual samples into one scalar representing quality of the whole database. Such a function would support identification of, for example, defective sensors, underperforming sites, and seasonal or secular trends. Slap quality addresses the problem of how to combine quality scores of each finger (i.e., right index, right middle,…) into one scalar representing quality of the slap fingerprints. This is relevant to DHS' operational needs with regard to US-VISIT's 10-print matching system.

In the paper titled *Performance of Biometric Quality Measures*, published in the April 2007 issue of IEEE Pattern Analysis and Machine Intelligence, NIST examined methods of assessing how effective a quality algorithm is in predicting performance. This activity supports future development of quality measurement algorithms since the ability to evaluate is necessary and vital during development.

NIST also conducted studies on incorporating quality in multimodal biometric systems and presented *When to Fuse Two Biometrics* at the Computer Vision and Pattern Recognition conference in June 2006.

Workshops

To discuss capabilities vis-à-vis operational requirements, and to identify research needs, testing requirements, and standardization gaps, NIST conducted a series of international Biometric Quality Workshops in March 2006 and November 2007. The workshops provided a forum for experts to share their research and discuss problems and new developments. Each workshop attracted more than 160 attendees to listen to more than 40 presentations of the world's leading technologies. The workshops are aimed at improving accuracy of biometric systems by incorporating quality assessment technologies into the sample acquisition process. They aim to assess current quality measurement capabilities and to identify technologies, factors, operational paradigms, and standards that can measurably improve quality.

f. Multimodal

i. MBGC

Over the last decade, numerous government and industry organizations have or are moving toward deploying automated biometric technologies to provide increased security for their systems and facilities. Results from the FRVT 2006 and FRGC documented two orders of magnitude improvement in the performance of face recognition under full-frontal, controlled conditions over the last 14 years. For the first time, ICE 2006 provided an independent assessment of multiple iris recognition algorithms on the same data set. However, further advances in these technologies are needed to meet the full range of operational requirements. Many of these requirements focus on biometric samples taken under less than ideal conditions, for example:

- low-quality still images;
- high- and low-quality video imagery;
- face and iris images taken under varying illumination conditions; and
- off-angle or occluded images.

Building on the challenge problem and evaluation paradigm of FRGC, FRVT 2006, ICE 2005, and ICE 2006, the Multiple Biometric Grand Challenge (MBGC) is designed to address these problem areas. One of the highlights of the MBGC is the Portal Challenge problem. The success of the iris at a distance project (HumanID) and the "Iris on the Move" System led to the design of the MBGC Portal Challenge problem. The goal of the Portal Challenge problem is to develop algorithms that recognize people from near-infrared image sequences and high-definition video sequences. The sequences will be acquired as people walk through a portal.

The MBGC started with a kick-off workshop on April 18, 2008, and plans to have the first set of results by early 2009. Plans are to follow the MBGC with an independent technology evaluation to verify MBGC results.

ii. MBARK

Despite existing efforts, building modern biometric applications (or *clients*) that are flexible with respect to changes in sensors, workflow, configuration, and responsiveness remains both difficult and costly. The Multimodal Biometric Application Resource Kit, or MBARK, reduces the complexity and costs of implementing such an application. MBARK is public domain source code that may be leveraged to develop the next-generation of biometric and personal identity verification applications.

Incorporating the MBARK libraries can yield a variety of enhancements critical for the success of any real-world system. For example, MBARK provides a usability-tested and consistent user interface. MBARK provides operators with the means to quickly recover from both minor mistakes and major hardware failures. In addition, the use of XML facilitates true sensor interoperability via plug-ins and allows for changes in workflow on-the-fly.

MBARK represents an immediate and field-tested response to The National Biometric Challenge of developing middleware techniques and standards that will permit "plug-and-play" capabilities for biometric sensors.

The following are just some of the features of MBARK that make it robust and flexible with respect to changes in sensors, workflow, configuration, and responsiveness.

- **Provides a consistent user interface**
  A user-centered and consistent user interface reduces errors and minimizes the need to retrain users as vendors develop new sensors and software. The benefits of usability-driven design are well understood.

- **Allows users to recover from mistakes**
  Significant costs may accompany any system that does not allow recovery from both common and uncommon mistakes. With MBARK, an operator can not only easily recover from mistakes, but may also save a snapshot of a session (in the form of an XML file) and load it again at a later time.

- **Adjusts workflow automatically**
  Defining a workflow that accommodates mistakes becomes more complex as "edge cases" are added. For example, how should the system behave if a fingerprint sensor detects that a finger is missing but the operator has not indicated such?

- **Responds to user input**
  Users expect modern applications to be responsive to their input at all times—during initialization, startup, capture, task editing, and so on. How does a user distinguish between a long-running operation and a system that is simply "frozen"? MBARK uses a natively multi-threaded architecture to allow as much "background" processing as possible.

- **Provides true sensor interoperability**
  MBARK uses a plug-in style mechanism that allows true sensor interoperability based on a unified API—a common interface that has been

used to successfully integrate real face cameras, fingerprint scanners, and iris sensors. The MBARK architecture allows new sensors to be deployed without the need to even restart an MBARK application.

- **Provides flexible user configuration**
  A highly configurable biometric client empowers users to define and experiment with various biometrics and workflows without depending on any particular vendor to implement such changes. With XML files, MBARK allows users to define precise custom workflows specifically tailored to their needs.

- **Open and free**
  MBARK source code is public domain—the benefits of free software are well-discussed elsewhere. The GNU document Categories of Free and Non-Free Software contains more information about the differences between *open source* and *public domain software*.

More information about MBARK may be found at the project's website, http://mbark.nist.gov.

g. CITeR

The Center for Identification Technology Research (CITeR) is a National Science Foundation (NSF) Industry/University Cooperative Research Center (I/UCRC). Initial discussions and planning for CITeR began in the late 1990s and was funded for its first five years of operation as an I/UCRC in December of 2001. CITeR was renewed after external peer review for a second five years of operation in December 2006.

CITeR works with its affiliates to advance identification technology through cooperative definition and completion of highly leveraged research, education of the next generation of scientists and engineers, and effective knowledge transfer.

The results of the center's highly leveraged research are formally disseminated to members at twice yearly meetings. In addition to reports, papers, and site visits, access to the student researchers as potential future employees represents perhaps the most important long-term means of knowledge transfer.

h. Biometrics Usability

A more recent avenue of scientific research is the human computer interaction (HCI) of biometric systems. HCI and usability guidelines were well established for desktop systems, applications, and web applications that allow developers to design systems according to HCI principles and established baselines. However, no such HCI guidelines existed for biometric systems. DHS recognized this need and initiated a program in 2004 with NIST to develop HCI guidelines and standards for biometric systems.

The goal of the usability effort is the development and testing of a set of usability guidelines for biometric systems that:
- enhance performance (throughput and quality);
- improve user satisfaction and acceptance; and
- provide consistency across biometric system user interfaces.

Achieving these goals requires an understanding of the users, user behavior, and the systems' usability.

Six usability research studies have been conducted, including the study of the impact of:
- user habituation or acclimatization;
- counter height and anthropometrics;
- instructional materials;
- adaptable devices for accessibility;
- international symbols;
- relationship of counter height and angle of fingerprint scanners; and
- face overlays.

These research studies have resulted in seven reports and two ISO standards submissions. These documents provide guidelines for implementation and deployment of biometric applications. The test results have had a direct impact on existing and planned biometric deployments within US-VISIT.

i. Standards

While some very successful biometric standards activities existed well before 2001, such as the ANSI/NIST ITL-1-200x standards, the BioAPI specification (released March 2000) developed under the BioAPI Consortium with participation of government agency representatives, and the Common Biometric Exchange Formats Framework (CBEFF) specification developed by a group lead by NIST and NSA, 9/11 provided an impetus to greatly expand and accelerate comprehensive standards development as envisioned government and private sector systems required a solid standards base. The first step was to create formal standards working technical groups in accredited, existing standards development organizations to develop generic biometric standards that would support both identification and verification applications. The U.S. government spearheaded this effort by formally proposing these groups at the national (INCITS – InterNational Committee for Information Technology Standards) and international Joint Technical Committee 1 of ISO/IEC levels in October of 2001 and January of 2002, respectively. To further support these efforts, the U.S. government also assigned personnel and fiscal resources to lead these efforts. Since that time, 22 national standards and 25 international standards have been developed and approved. Several of these standards are now in their second versions.

Additional government-based standards activities include the following.
- Common Biometric Exchange Formats Framework (CBEFF), NISTIR 6529-A, April 2004 (proposed as an American National Standard, approved in 2005

as ANSI INCITS 398-2005. Since then a revision of this standard was approved in 2008 as ANSI INCITS 398-2008.

- Release of beta versions of BioAPI Conformance Test Suites (CTS), September 2005 by NIST and DoD.
- A Taxonomy of Definitions for Usability Studies in Biometrics, NISTIR 7378, November 2006.
- Approval of ANSI/NIST-ITL 1-2007, Data Format for the Interchange of Fingerprint, Facial and Other Biometric Information – Part 1, April 2007.
- Conformance Testing Architecture and Test Suite for data instantiations of CBEFF (ANSI INCITS 398-2008) developed by NIST and made available to the U.S. government – June 2008.
- Initiation of the development by NIST of Conformance Test Suites for biometric data interchange formats started June 2008.
- Initiation of a second version of the ANSI/NIST-ITL standard for an XML format. ANSI/NIST-ITL 2-200X, Data Format for the Interchange of Fingerprint, Facial and Other Biometric Information – Part 2: XML, which is undergoing ballot and public review in the summer of 2008.

By 2007, multiple competing versions of some standards existed[28]. To help ensure interoperability of government systems, the NSTC Subcommittee on Biometrics and Identity Management led an interagency effort to develop the *NSTC Policy for Enabling the Development, Adoption and Use of Biometric Standards* (September 2007). The goal of this policy is to establish a framework to reach interagency consensus on biometric standards adoption for the federal government and resulted in the release of the *Registry of US Government Recommended Biometric Standards* (July 2008). Federal agency adoption of these recommended standards and associated conformity assessment programs will enable necessary next generation federal biometric systems, facilitate biometric system interoperability, and enhance the effectiveness of biometrics products and processes.

---

[28] This is not uncommon in the standards realm, as the national and international standards bodies work at a different pace.

Appendix B - About the NSTC Subcommittee on Biometrics and Identity Management

The NSTC is the principal means within the Executive Branch to coordinate S&T policy across the diverse entities that make up the federal research and development enterprise. Chaired by the President, the membership of the NSTC is made up of the Vice President, the Director of the OSTP, Cabinet Secretaries and agency heads with significant S&T responsibilities, and other White House officials. The Subcommittee on Biometrics and Identity Management serves as part of the internal deliberative process of the NSTC. Reporting to and directed by the Committee on Technology, the Subcommittee's tasks are to:

- For Biometrics:
    - o Provide technical leadership in the development and implementation of interoperable federal biometric systems;
    - o Develop and implement multi-agency investment strategies that advance biometric sciences to meet public and private needs;
    - o Develop and adopt biometric standards as specified in the NSTC Policy for Enabling the Development, Adoption and Use of Biometric Standards;
    - o Develop consensus strategic outreach plans for biometrics, including collaboration on www.biometrics.gov, the annual Biometric Consortium Conference and other events;
- For Identity Management (of which biometrics is a subset):
    - o Identify cross-sector IdM issues and develop and implement plans to address the federal government's priority S&T needs;
    - o Facilitate the inclusion of privacy-protecting principles in IdM system design;
    - o Promote a scientifically educated and aware public that properly understands IdM technologies, federal programs, and issues;
    - o Strengthen international and public sector partnerships to foster the advancement of IdM technologies.

Co-chairs
Russell Neuman, OSTP (2002-2003)
Gary Strong, DHS S&T (2003-2004)
Kevin Hurst, OSTP (2003-2005)
Duane Blackburn, FBI, OSTP (2004-Present)
Brad Wing, DHS/US-VISIT (2006-2008)
Chris Miles, DOJ/NIJ, DHS (2006-2007, 2008)
James Dray, DOC/NIST (2007-Present)

Executive Secretaries
Karen Walker, DHS/S&T (2003-2005)
Kim Shepard, DOJ/FBI/SETA (2005-2007)
Michelle Johnson, DOJ/FBI/SETA (2007-Present)

Subcommittee Participants[29]

Donald Anderson, EOP/NSC
Joseph Arata, DoD
Douglas Arnold, DoD
Joanne Arzt, State
John Atkins, State
Carol Bales, EOP/OMB
Bill Baron, DOT/RITA *
Sankar Basu, NSF*
Andy Black, DOJ/BOP
Duane Blackburn, DoD, DOJ/FBI, EOP/OSTP *
Janice Bland, DOJ/FBI
Janet Boodro, DOJ/JMD
John Boyd, DoD/Navy
Sam Bradley, DOE
Chris Brazier, State
Fred Bunke, DHS
Susan Burk, State
Michael Butler, GSA
Zaida Candelario, Treasury/IRS
Sam Cava, DoD, DOJ/FBI*
Michael Chang, NCTC
Ed Chase, EOP/OMB
Patricia Cogswell, DHS/SCO
Greg Collett, DHS/CIS
John Cook, State, NCC
Maureen Cooney, DHS/Privacy
Thomas Coty, DHS/S&T
Bert Coursey, DOC/NIST
Dan Cundiff, DoD/AT&L
David Cuthbertson, DOJ/FBI
Tim Daugherty, DHS/USSS
Aaron Davenport, EOP/OVP
Jeff David, DoD/CTTSO
Tom Dee, DoD/AT&L *
Kimberly Del Greco, DOJ/FBI *
Semahat Demir, NSF
Stephen Dennis, DHS/S&T
Trent DePersia, DOJ/NIJ, DHS/S&T
MaryBeth Dormuth, DOT/FAA
Jim Dray, DOC/NIST*
Jeff Dunn, NSA*
Stanley Erickson, DOJ/NIJ
Travis Farris, State/CA
Sarah Francia, State
Art Friedman, DoD
Cita Furlani, DOC/NIST
Jim Zok, DOT/MARAD*

Michael Garris, DOC/NIST
Alexandra Gianinno, EOP/OMB
Bill Gravell, DoD/Navy
Myra Gray, DoD/BTF
David Grazer, EOP/NSC
Patrick Grother, DOC/NIST
Joseph Guzman, DoD/BMO*
Ed Harras, DOT/FAA
Monte Hawkins, NCTC, EOP/HSC
Martin Herman, DOC/NIST
David Herrington, DoD/CTTSO
Douglas Hill, State/CA
Mike Hogan, Commerce/NIST*
Robert Holman, DOJ/FBI
Thomas Hopper, DOJ/FBI
John Hotchner, State/CA
Kevin Hurst, EOP/OSTP
Larry Jellen, GPO*
Usha Karne, SSA
Kim Keefer, DHS/OCRCL
Joe Kielman, DHS/S&T
Michael King, ITIC, IARPA*
Nuala Kelly, DHS/Privacy*
Andy Kirby, ITIC*
Eva Kleederman, ODNI
Kathleen Kraninger, DHS/SCO
Scott Lamoreux, DOJ/FBI
Jim Lantzy, NCTC
Neal Latta, DHS/US-VISIT, TSC
Rick Larazrick, DOT/FAA, DHS/TSA*
Michael Lilienthal, DoD/BTF
Valerie Lively, DHS/S&T*
Dave Lohman, DoD/BTF
Philip Loranger, DOT/OCIO
James Loudermilk, DOJ/FBI
Jim Mahan, DOJ/FBI
Joseph Maher, DHS/OCRCL
Richard Martin, State/CA
Michael McCabe, DOC/NIST
Ed McCallum, DoD/CTTSO
Thomas McKenna, DoD/ONR
Ross Michaels, DOC/NIST*
Chris Miles, DOJ/NIJ, DHS/S&T*
Kenneth Mortensen, DHS/Privacy, DOJ/Privacy
Frank Moss, State/CA

Greg Motta, DOJ/FBI
Todd Mullenax, DOJ/FBI
Michael Neifach, EOP/HSC
Elaine Newton, DOC/NIST
Kirstjen Nielsen, EOP/HSC
Joyce Nyman, DHS/CG
Omid Omidvar, DOC/NIST*
Eric Panketh, EOP/OMB
Jonathon Phillips, DoD/DARPA, DOC/NIST*
Richard Phillips, Treasury/IRS
Jennie Plante, DOJ/USA
Carl Pocratsky, DOE
Fernando Podio, DOC/NIST*
Troy Potter, DHS/US-VISIT
Niels Quist, DOJ/Privacy*
Ben Riley, DoD/AT&L
Bob Ross, DOT/OGC
Peter Sand, DHS/Privacy*
Jeff Sarnacki, EOP/HSC
George Saymon, DHS/FBI
Marie Sciarrone, EOP/HSC
Susan Sexton, DOT/FAA
Paul Shannon, EOP/HSC
Kristen Sheldon, EOP/HSC
Nicole Spaun, DOJ/FBI
Richard Stewart, DOT/NHTSA
Scott Swann, DOJ/FBI, CJIS
Barbra Symonds, DOT/IRS
Elham Tabossi, DOC/NIST
David Temoshok, GSA
Gerald Thames, NCTC*
Mary Theofanos, DOC/NIST
Jeanette Thornton, EOP/OMB
Fred Vogel, State
Richard Vorder Bruegge, DOJ/FBI
Karen Walker, DHS/S&T
Kimberly Weissman, DHS/US-VISIT*
Kamela White, EOP/OMB
Charlie Wilson, DOC/NIST
Diane Wilson, DHS/TSA
James Windle, EOP/OMB
Brad Wing, DHS/US-VISIT, DOC/NIST*
John Woodward, DoD/BMO*
Steve Yonkers, DHS/US-VISIT

---

[29] Since 2002; those that have served in an interagency leadership role within the Subcommittee at some point it its existence are noted with an asterisk (*).

www.ingramcontent.com/pod-product-compliance
Lightning Source LLC
Chambersburg PA
CBHW041234200526
45159CB00031B/1207